BUILDING
FIREPLACE MANTELS

BUILDING FIREPLACE MANTELS

*Distinctive Projects for
Any Style Home*

MARIO RODRIGUEZ

The Taunton Press

 The Taunton Press
Inspiration for hands-on living™

The Taunton Press, Inc., 63 South Main Street, PO Box 5506,
Newtown, CT 06470-5506
e-mail: tp@taunton.com
Distributed by Publishers Group West

EDITOR: Tony O'Malley
COVER DESIGN: Steve Hughes
INTERIOR DESIGN: Lori Wendin
LAYOUT: Suzie Yannes
ILLUSTRATOR: Ron Carboni
PHOTOGRAPHERS: Bruce Buck, Mario Rodriguez

LIBRARY OF CONGRESS CATALOGING-IN-PUBLICATION DATA
Rodriguez, Mario, 1950–
    Building fireplace mantels : distinctive projects for any style home / Mario Rodriguez.
        p. cm.
    ISBN 1-56158-385-5
    1. Mantels. I. Title.
    TH2288 .R63 2002
    684.1'6--dc21

                                                    2002007107

Printed in the United States of America
10 9 8 7 6 5 4 3 2 1

The following brand names/manufacturers are trademarks: Masonite, Phenoseal, and Rust-Oleum.

ABOUT YOUR SAFETY: Working with wood is inherently dangerous. Using hand or power tools improperly or ignoring safety practices can lead to permanent injury or even death. Don't try to perform operations you learn about here (or elsewhere) unless you're certain they are safe for you. If something about an operation doesn't feel right, don't do it. Look for another way. We want you to enjoy the craft, so please keep safety foremost in your mind whenever you're in the shop.

To my son Pete, a budding woodworker.

ACKNOWLEDGMENTS

I'd like to thank my students at the Fashion Institute of Technology: Les Katz, Charlie James, Lorraine Servino, Alicja Patrzalek, Gina Balzano, Dana Sawyer, Michael Scarborough, Edward Fitzpatrick III, and my studio technician, Patrick Stuzinski, who helped build, finish, photograph, transport, and install the fireplace mantels.

My appreciation to Lauren Farber and Steve Dworkin, Richard Nolan, Judy Rusignuolo, Ann Miller, Dr. Hugh Crean, and the Restoration Department at the Fashion Institute of Technology, NYC, for their generous support and the loan of space, various props, rugs, plants, and furniture.

At Taunton Press, I'd like to thank Helen Albert, and Carolyn Mandarano and my editors, Strother Purdy, Tom Clark, and Tony O'Malley. And finally, my photographer, Bruce Buck.

CONTENTS

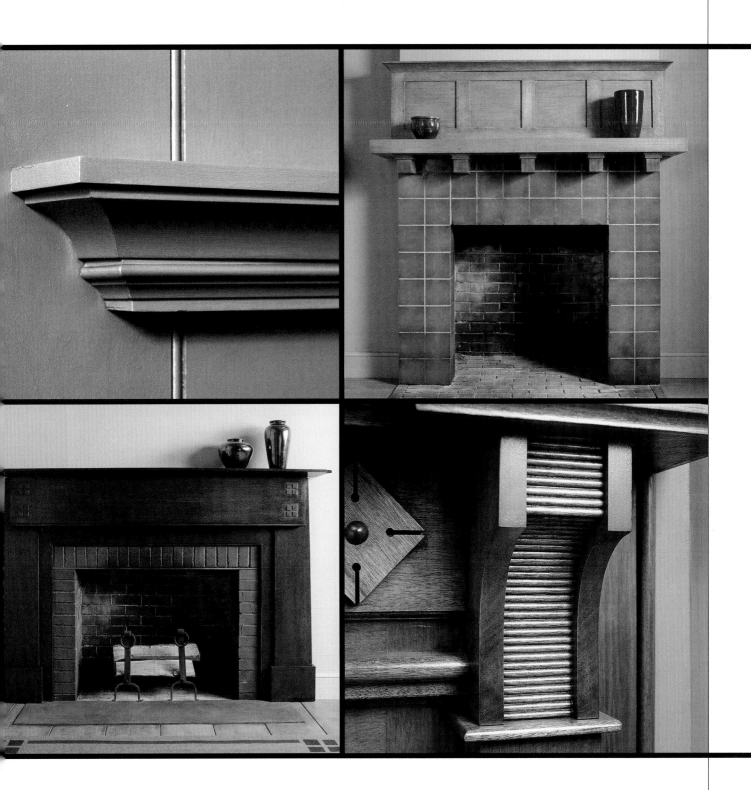

INTRODUCTION

The idea for this book came from the steady stream of comments and questions generated by the material on fireplace mantels I've written for Taunton Press over the years. Of all the woodworking projects I've built, and written about, the fireplace mantel projects have drawn the strongest- and longest-lasting response.

I think that even for owners of contemporary homes, equipped with powerful and efficient heating systems, the fireplace has come to symbolize a safe and stable refuge from a hectic and sometimes hostile world. Commonly situated in the family or living room, it is regarded as the hub or center of the traditional home and often serves as an attractive backdrop for both solemn and festive family occasions. So, even if it's not used to provide heat, the design and decoration of the hearth is considered an essential part of an attractive home.

By presenting a broad range of designs, this book contains something for almost everyone—and every style interior. I've designed and built mantels dating from the early 18th right up to the 20th centuries. There are several early American designs for traditionalists, a Victorian mantel stuffed with rosettes and bold trim, a solid but elegant Arts and Crafts design, and even a zigzag Art Deco piece. Altogether, the projects in this book present a broad range of mantels, not only in various styles, but in different skill levels, too.

Like my previous book, *Traditional Woodwork,* my goal was to present well-designed projects, built with time-saving techniques and modern materials available to the average woodworker. Almost every mantel features a technique or product that could be useful in the design and construction of woodworking projects other than fireplace mantels. And while designing and building the mantels, I tried to keep one goal in mind: The finished product should definitely appear more difficult and complicated to build than it proved to be during construction.

While writing this book and building the mantels, I was fortunate to have students from the Restoration Department of the Fashion Institute of Technology, eager to help. It was tremendously rewarding to work side by side with my students. The experience allowed me to measure how well I taught them and to gauge the clarity of my material (for the book). I was happy to discover they also made excellent models, posing and contorting their bodies while I struggled to get the perfect shot.

MANTEL-MAKING BASICS

THE MANTEL IS a purely decorative frame for a fireplace. A beautiful mantel may add to the character of the room and enhance your experience of sitting by the fire, but it has no effect on the working characteristics of the fireplace itself. Nevertheless, before you set out to make a mantel for your fireplace, it's worth taking the time to understand how a fireplace works.

ANATOMY OF A FIREPLACE

A fireplace is not simply a hole cut into a wall to accommodate a fire. There are various considerations that bear upon its design and construction. A good working fireplace is well insulated from other combustibles in the room, thoroughly burns its fuel, projects heat efficiently into the room, and evacuates any

MANTELS AND BUILDING CODES

Each fireplace mantel in this book was built to be used on a real working fireplace. Most of the designs can be adjusted in size to fit your fireplace opening. Bear in mind that there are building codes that regulate certain conditions and materials of fireplace mantel construction. Most relevant are the setback distances from the edge of the firebox opening to the combustible mantel material. In the case of a brick firebox, the setback may be defined by the brickwork. But it's worth checking regardless of the material. The last thing you want is for your mantel to be damaged by an overly robust fire.

smoke up through the chimney. That's a tall order, and any of those requirements not fulfilled can become inconvenient or even dangerous. Let's examine some of the factors that are essential to good fireplace design and performance.

The hearth opening is the actual opening in the fireplace wall. Too small an opening can mean an undersized fire that's unable to provide adequate heat for the room or any visual pleasure. On the other hand, a large opening will require a bigger fire in order to properly burn. This could result in an uncomfortable room, either too warm or too cool, and wasted wood. A height of 30 in. to 36 in. and a width of 42 in. to 48 in. are considered ideal dimensions for the average residential room.

The firebox contains the fire and can be built of either brick or stone.

The side jambs are the side walls of the firebox. Splayed outward, they help to radiate heat into the room.

The back wall is made of the same material as the side walls. It was sometimes decorated with a cast-iron fire back that increased the heat radiated into the room.

The lintel is the structural member that spans the opening and supports the masonry above. Surprisingly, in some early homes, the lintel was a stout oak beam that stretched across the fireplace opening.

The back hearth is the floor of the firebox, closest to the back wall. Today, this is often

made of brick. You might find fieldstone used in some older homes.

The front or extended hearth is the portion of the floor in front of the firebox extending into the room and past the ends of the opening. This nonflammable deck protects the flammable sections of the floor surrounding the hearth from sparks. The front hearth can be any hard nonflammable material,

ANATOMY OF A FIREPLACE

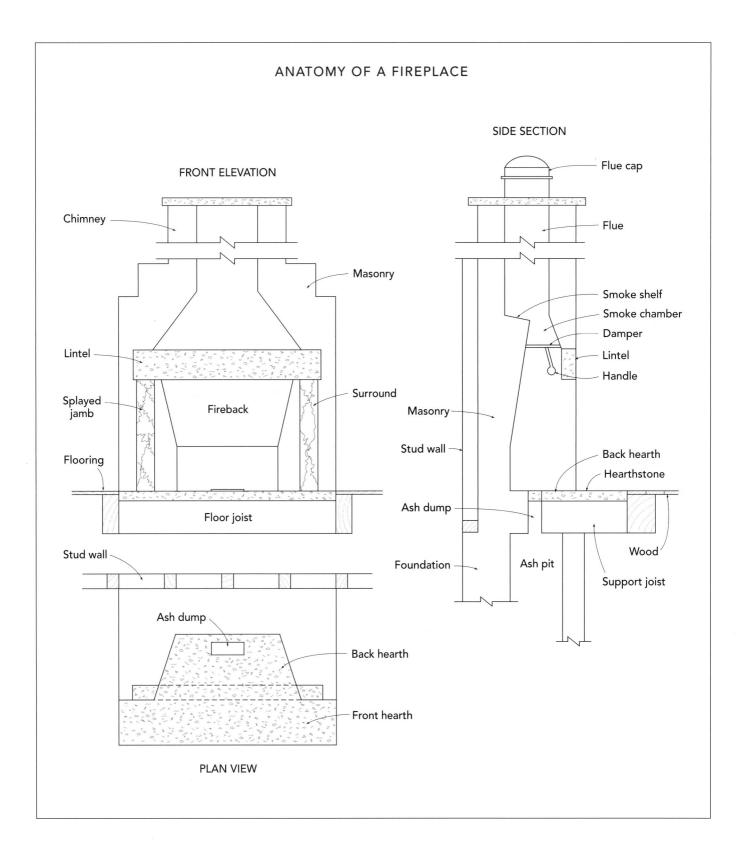

FRONT ELEVATION

Chimney

Masonry

Lintel

Splayed jamb

Surround

Fireback

Flooring

Floor joist

Stud wall

Ash dump

Back hearth

Front hearth

PLAN VIEW

SIDE SECTION

Flue cap

Flue

Smoke shelf

Smoke chamber

Damper

Lintel

Handle

Masonry

Stud wall

Back hearth

Hearthstone

Ash dump

Foundation

Ash pit

Wood

Support joist

such as brick, fieldstone, concrete, tile, or quarried stone.

The throat is the passage through the masonry stack that allows the evacuation of the smoke from the firebox.

The damper is a movable metal flap placed at the top of the firebox. It's used to control the rate of combustion (how fast or slow a fire burns) and, when the fireplace is out of service, prevents warm room air from escaping up the chimney.

The smoke chamber is a space between the top of the throat and the bottom of the flue. With sloping sides, it funnels the smoke up the flue.

The smoke shelf is an angled shelf that diverts downdrafts from the chimney and prevents smoke from blowing back into the room.

The flue is a smooth-walled tube that removes smoke from the home. In common fireplace construction, it is frequently made up of square clay sections, joined end to end. Today, some modern flues are made of circular metal sections joined together.

The chimney is the hollow masonry stack that surrounds the flue and extends above the roofline of the house.

The flue cap covers the flue opening. This lid prevents downdrafts and rain from entering the flue opening. In addition, some caps have a mesh screen that prevents dangerous sparks from escaping up through the chimney and keeps birds, bats, and squirrels from entering.

The surround is the area around the face or opening of the fireplace. In conventional fireplace construction, it's made of noncombustible material, such as marble, tile, stone, or brick. For safety, experts recommend that any combustible materials, such as wood, be set back a minimum of 8 in. from the fireplace opening.

The ash dump is an opening in the floor of the firebox that leads to a compartment in the crawl space or cellar for the ashes. This eliminates the unpleasant task of cleaning out the hearth with a broom and shovel. Because of the possibility that live embers might pass from the hearth, the ash dump container, lids, grates, and other accessories must be made of noncombustible materials.

ANATOMY OF A MANTEL

The fireplace mantel provides no structural support to the fireplace. It doesn't improve the burning of a fire or contribute to safety. It is simply decorative. For the mantelpiece to contribute to the décor of the room, it must exhibit grace, beauty, and good proportions. It must blend in with other architectural features in the room, yet stand out as a focal point. Just as every house is a set of component parts, most mantels share a list of parts that can be identified individually:

Plinth is a projecting block resting on the floor at the base of a column or pilaster. It is usually chamfered or molded at the top.

Pilaster is a rectangular projection, similar to a column, that's attached to the wall. On a fireplace, pilasters flank the opening and typically sit atop plinth blocks.

Abacus is a molding or separation between the top of the pilasters or columns and the entablature.

Entablature is the entire superstructure of moldings above the abacus.

Frieze is the middle band between the cornice and abacus molding that spans the fireplace opening.

Cornice is the projecting molding that crowns an entablature.

Mantel shelf is the top horizontal surface of the mantel that extends out beyond the cornice. Sometimes the contour of the mantel shelf mirrors in small scale the outline of the mantel components below, taken as a whole.

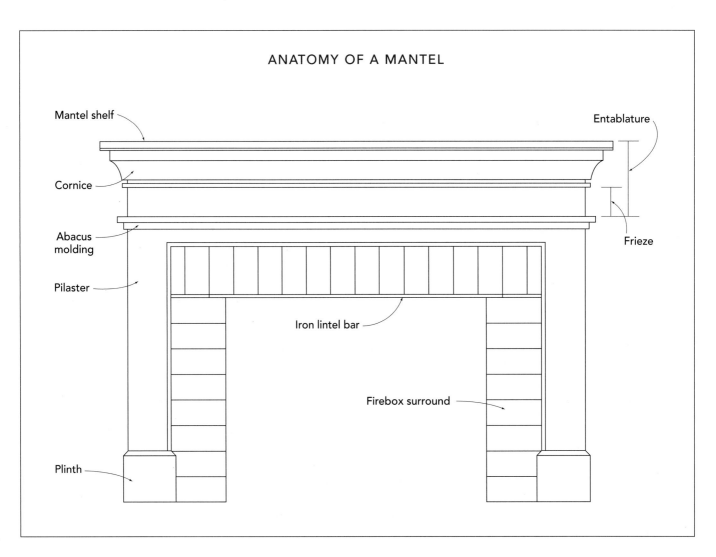

ANATOMY OF A MANTEL

Mantel shelf

Entablature

Cornice

Frieze

Abacus molding

Pilaster

Iron lintel bar

Firebox surround

Plinth

How a Fireplace Works

No matter what you burn or where you burn it, a fire needs oxygen from the air to sustain combustion. Once the fuel ignites and becomes a flame, the air above the fire is warmed, creating a rising current of warm air and pulling new air from the room into the fire. This is basically what occurs in a fireplace. But aspects of the fireplace's design, construction, and operation should make effective use of the principles of combustion.

Let's take a closer look at this as it occurs inside the fireplace. When the fuel ignites into a flame, it creates a current of warm air shrouded in smoke. But before a strong enough current is formed to draw the smoke up and away, it lingers and swirls in the fireplace's throat and smoke chamber. When the flue and smoke chamber are sufficiently heated to draw away any smoke, the smoke shelf deflects any cold air traveling down the chimney, and prevents any smoke lingering in the smoke chamber from being blown back into the room. Eventually, the fire will heat the smoke chamber and flue and draw some of the warm air and any smoke away from the room.

Once a strong fire is burning, the heated firebox, smoke chamber, and flue will create an updraft that will support a hot, smoke-free fire in the hearth.

Mantel-Making Materials and Techniques

Most of the early 17th- and 18th-century American mantels were made of solid wood. Even in the 19th century, the popularity of wood continued to exceed that of other materials.

With wood, a craftsman can design and build almost any style, size, or color mantel. Wood can be shaped, bent, laminated, carved, turned, or veneered to produce an endless variety of details. Another advantage of using wood is the enormous range of colors and textures it comes in.

There are basically three wood materials available today that give the modern craftsperson great range, in terms of design and construction techniques, at an affordable price. Each of these popular and readily available wood materials are sold through local suppliers, are easily transportable in the back of the family mini-van, and can be cut, shaped, and decorated with common home shop tools.

Solid wood

Several of the mantels in this book are based on early examples, originally made of solid wood. Two centuries ago this material was readily available in wide, thick boards. And by using solid wood to build his mantel, the early craftsman could achieve speed, uncomplicated design, and honest construction. Most of the early 17th- and 18th-century American mantels were made of solid pine.

Later, during the 19th century, solid hardwoods were used for architectural woodwork, including fireplace mantels. In many instances, hardwood was employed for its striking grain patterns and rich color. The durability of hardwood was also a factor. Hardwood could better survive the accidental bumps and bruises that were a consequence of daily life. And since, by then, power machinery was available, the material's hardness didn't pose a problem in the production of moldings, doors, or mantels.

Today, solid wood is still the overwhelming favorite. It can be carved or recut to yield book-matched (mirror image) panels, and lam-

inated into lively twisting shapes or curves while retaining the appearance of solid wood, or even turned on a lathe.

Plywood

This material is perfect for building mantels, offering several advantages over solid wood boards. Foremost, since it is available in 4 by 8 sheets, there is no need to glue up several narrow boards to cover a large area. Another advantage is stability. Because of its multiple alternating layers, the material is stable and unaffected by quickly changing temperature or moisture conditions. And, finally, plywood is available in a range of beautiful veneers that can provide an appearance as fine as any free-standing chest or table. In my designs, I take advantage of its stability and combine it with solid wood edges to hide the laminated core. This technique gives me a durable repairable edge that can also be shaped. The combination of plywood and solid edging gives me the con-

venience and economy of sheet goods and the good looks of solid wood.

Medium-density fiberboard (MDF)

Like plywood, medium-density fiberboard comes in 4 by 8 sheets. This dense gray-brown material is actually pulverized wood fibers that have been heated and compressed after resins have been added to improve strength and stability.

For low-budget mantels that will be painted, MDF could prove useful. This material can be cut and shaped just like wood. But unlike wood, it has no end grain. And once sealed and primed, it takes a beautiful coat of paint.

A major problem with MDF is durability. Because of its constitution, MDF is susceptible to bumps and physical damage. Another disadvantage is its weight; it is extremely heavy, yet it sags under its own weight. And in the event of contact with water, MDF can swell

FIREPLACE EQUIPMENT

There is some equipment that I've found useful for the safe operation and enjoyment of a fireplace. None of it is particularly difficult or expensive to obtain. The purpose of these accessories is to promote safe and efficient burning of material in the fireplace and enhance your enjoyment of the fireplace.

Andirons are primarily decorative and are used to raise the logs off the hearth floor in order to promote thorough burning. But as the logs burn through, they fall between the andirons.

Grates give more thorough support of the logs and smaller pieces of firewood. This is a more utilitarian piece of hardware; although not as fancy as andirons, it performs the same function.

Screens provide protection from airborne sparks and embers. They are available as flat one-piece panels supported by feet, or as three-part units joined by hinges. Another variation is the curtain or drape that can be pulled across the fireplace opening.

Fireplace tools are available as sets and contain a shovel, tongs, a broom, and a poker. These tools allow you to move and rearrange the burning logs and afterward to sweep out the hearth.

and distort in thickness. This distortion can cause any applied coating to peel, and fasteners to come loose.

TECHNIQUES FOR JOINING WOOD COMPONENTS

Screws

Screws offer the most holding power for their size and the effort required to install them. When properly applied (correct length and size), they will hold forever. Yet they can be easily removed without causing any damage to the workpiece. I find that to take full advantage of their unique capabilities, screws are best set into predrilled countersunk holes. That way their full holding power is employed without any risk of splitting the wood and damaging the project. Another advantage of using screws is that large projects that will ultimately require on-site installation can be fully assembled in the shop, then taken apart, and later reassembled on the job site.

Screws come in a variety of metals, finishes, sizes, lengths, and thread patterns. For interior architectural work, I favor #8, stainless steel, wide-thread screws, with a #2 Phillips head.

This type of screw is easy to drive, doesn't rust, and because of the wide thread pattern, can be installed and removed quickly.

Wire nails and finish nails

Nails provide the oldest and simplest way to join two pieces of wood. They are inexpensive, easy to obtain, and easy to use. They come in a variety of sizes to suit the need. For building mantels you'll commonly use fourpenny, six-penny, and eightpenny nails. I would recommend them in situations where speed is essential. A hammer and a nail set are the only tools you need to install them. And in most applications, they provide adequate holding power. Be careful driving the nails so you don't ding up the work. Stop driving with the nail just proud of the wood, and use a small nail set to drive them to about $\frac{1}{16}$ in. below the surface.

Air-powered staple and nail guns

Pneumatic nailers and staplers can't be beat for speed, convenience, and ease of use. One squeeze of the trigger shoots and countersinks the fastener in less than a second. And you can position the workpiece with one hand while firing the nail gun with the other. But the best thing about nail guns is a blemish-free

Pneumatic nailers and staplers make quick work of driving fasteners, and you can hold the work with one hand while the nailer sinks the nail.

job. There are no split moldings or hammer marks to repair or disguise later. The initial expense might discourage you, but these days every major tool manufacturer is producing a line of pneumatic fasteners and portable air-compressors at prices accessible to most woodworkers.

I own three types of guns: a mini-pin nailer, a finish nailer, and a narrow-crown stapler. The mini-pin nailer shoots tiny 21-gauge

SELECTING AND STORING FIREWOOD

For a safe, comfortable, and efficient fire, your firewood should be dry and seasoned. Normally, it takes from six months to a year for wood to achieve the proper moisture level. A visual inspection is the best way to determine whether your wood is ready. Just check for small radial cracks at the ends of the logs, and pry off a section of bark along the center of the log to check whether the wood underneath is dry to the touch. Purchased in late winter or early spring and stacked and covered properly, firewood should be ready to burn by winter.

Firewood should be stored neatly on a rack that promotes ventilation and drying. Make sure that the bottom of the rack is off the ground and that the ends are properly braced—a tumbling woodpile is unattractive and can be dangerous. And be sure to cover the rack with loose plywood panels or a tarp.

Make sure that each log segment is oriented to shed water. The rack itself should be lodged against a vertical anchor, such as a tree or a masonry foundation wall. But don't set your rack against any wooden building structure; I believe this practice can promote the migration of wood parasites or rodents into the building.

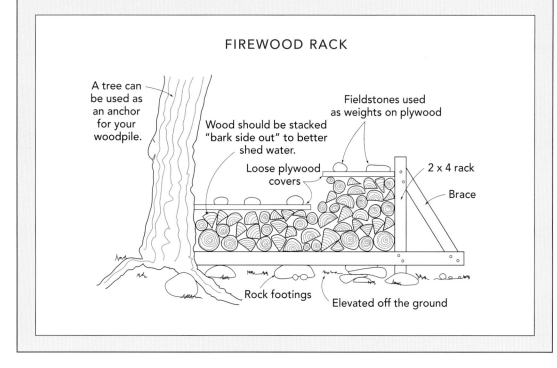

FIREWOOD RACK

A tree can be used as an anchor for your woodpile.

Wood should be stacked "bark side out" to better shed water.

Fieldstones used as weights on plywood

Loose plywood covers

2 x 4 rack

Brace

Rock footings

Elevated off the ground

headless pins that are perfect for attaching small narrow moldings and aligning the corners of miters. And the small hole created by the tiny pin is easily filled with a smear of wax or filler. The finish nailer fires small 18-gauge nails ranging in length from ⅜ in. to 1⅜ in., and I use it in situations that require greater holding power. The narrow-crown stapler shoots staples from ½ in. to 1⅜ in. The term "narrow-crown" refers to the width of the staple and the narrow entry hole it makes. These staples have good holding power because the prongs converge when driven into the workpiece. This tool is great for joining plywood cabinet parts and attaching cleats and cabinet backs.

Tongue-and-groove or spline joints

This common joint can be made on the table saw or with a router. For the tongue-and-groove joint, the easiest and best-fitting method is to use matched router bits. This means that each half of the joint, shaped with a separate cutter, will fit its mate perfectly.

A variation on the tongue-and-groove is the spline joint. Both edges are grooved with the same router or dado bit, then the grooves are joined together with a separate spline.

TOOLS FOR MAKING MANTELS

Router

Used for everything from joinery to delicate edge decoration, the router is the most versatile of the modern woodworking machines. It can be used in a hand-held mode for edge shaping and simple joinery tasks, or it can be installed in a router table to extend its capabilities. Many of the mantel projects here use a router for making the decorative moldings and for tasks such as cutting rabbets or grooves for joinery.

Router table

A router table is a flat surface, fitted with an adjustable fence, that supports an upside-down router with the router's bit or cutter projecting up through a hole in the table. Material is pressed against the fence as it passes the cutter, forming the profile on the material. Using a router table will tremendously speed up your work and improve the quality of your moldings.

The sequence for producing moldings from wide boards is straightforward: First rout both

The router is one of the most versatile tools used in making mantels. Router bits come in hundreds of shapes to make lengths of molding for decorative details on mantels.

The router table offers the safest way to mill your own moldings for making mantels. It's also useful for router joinery operations.

edges of the board, cut the molding free on the table saw, then return the board to the router table and repeat the sequence. Sometimes the edge of a board will bow after ripping a piece off, so I sight down the edge and, if it's not straight, I'll run it over the jointer. I always produce about 20 percent extra linear molding, allowing enough material so that sections with router blips, tearout, and other blemishes can be discarded.

Biscuit joiner

There is probably no single tool that has had greater impact upon woodworking in the last 20 years than the biscuit joiner, sometimes called a plate joiner. Originating in Switzerland, this portable machine has swept over the United States. I can't think of a single serious cabinet-maker or furniture maker who doesn't own one.

Although there are several designs available, they all cut a shallow semicircular kerf into the wood. The location of the slot or kerf is registered using the base of the machine or a movable fence. When two mated kerfs are fit-ted with a compressed beechwood biscuit or plate (they resemble wood wafers) and glued, you have a tight and perfect joint. Once the joint is assembled and clamped, the com-pressed biscuit absorbs moisture from the glue and swells to hold the joint tight. The biscuit joiner virtually eliminates tedious corrections and cleanup work later.

Wood plane

An indispensable tool used in the building and installation of fireplace mantels is the wood plane. Primarily used to smooth and flatten wood, you'll reach for them time and again during construction for various tasks. Although there are several to choose from, I would recommend starting with a jack plane and a block plane.

Bench plane

I find the #5¼ jack plane is best for general bench work. With its long length (11½ in.), it will easily prepare joints and remove mill marks (faint scratches, ridges, and other blem-ishes incurred during the milling process).

The biscuit joiner uses football-shaped wafers to join wood components, providing an efficient and versatile modern substitute for many tradi-tional joinery techniques.

The Stanley #5¼ jack plane and the Lie-Nielsen #102 block plane came in handy when I was fitting mantel parts together and installing com-pleted projects.

Block plane

For all the final fitting and trimming tasks that are part of the installation, nothing beats a block plane. Handy because of its small size, it can be used accurately with one hand while you steady the work with the other.

BUYING AND MAKING MOLDINGS

Most of the mantels in this book make use of applied moldings. Sometimes the moldings can be bought off the shelf in a home center, while in other cases, because of the shape or the wood species involved, you'll need to make your own.

Buying molding

A wide range of molding profiles are sold in various materials: clear pine, oak, pre-painted, etc. Especially for painted mantels, you can usually find what you need at a home center. Molding stock should be clean and free of any discoloration. Dirty stock indicates material that's been around awhile. Also make sure your selection is a readily available profile—

just in case you need more. It's best to buy as much as you'll need at one time so the profiles will match up.

There is also a new generation of molding products available today made of composite materials. Especially for ornate crown moldings and the like, these are generally far less expensive than the same product would be if made of wood. The Georgian Mantel in chapter 3 makes use of both composite moldings and decorative applied ornaments that look just like exquisitely carved details.

Making moldings

The first step in producing molding in the shop is to joint and thickness all the material in order to ensure the uniformity of the completed molding stock. You should start any project like this with your stock flat and square. This is a critical step, necessary to guarantee that the moldings will match up and mate properly at mitered corners later.

For speed, accuracy, and safety, I leave my milled molding stock as wide as possible. Handling a wide board gives more stability to the routing operation and keeps my fingers safely out of the way. But at the same time, I cut my stock as short as possible. This makes it easier to keep the stock flat upon the router table or saw and against the fence.

Moldings shaped on the edge of a board can be cut free and used as edging on plywood, or applied to a face panel.

A router with a carbide-tipped round-nose bit was used to make this fluted pilaster.

To make moldings from wider boards, first rout the shape onto one or both edges of the board. Then rip the shaped edge from the board. Check that the new edge of the board is straight (run it over the jointer if it's not), then repeat the process.

Linear moldings can be used individually or combined with other moldings for more complex designs. They can also be cut into small segments, then reconfigured into a decorative pattern. They're commonly added to plywood panels to conceal the plywood core. Finally, even wide boards can be shaped to produce decorative elements, like fluted or reeded pilasters, a classical feature found on many fireplace mantels.

INSTALLING A MANTEL

Most mantels are built in the shop and then transported to the house site for installation. In the shop, a thorough woodworker takes into account every variable or condition that might be on the job site. If it's an old house and built before strict codes, you never know what's really behind the wall. So you have to be ready for anything.

Considering that the chimney wall is made of masonry, it's certain that some kind of masonry fastener will be needed. The trick is to attach the mantel to the masonry wall without disfiguring the finished mantelpiece.

Below are three fastening methods that meet this criterion:

Lead anchors and screws involve drilling a hole through the wall and into the masonry with a carbide-tipped drill bit and then placing a lead sleeve in the hole. The sleeve expands with the insertion of the screw and holds tight.

Masonry screws require a predrilled hole in an exact size to match the screw thread, but an insert is not used. Instead the hardened threads of the screw cut into the masonry and hold securely. Additionally, masonry screws can be removed for repositioning of the workpiece, then driven in again.

Construction adhesive is dispensed from a tube and applied with a caulking gun. This material has minor gap-filling properties, sets in a few hours, and bonds wood to almost any other material.

Attaching the foundation of a mantel directly to the wall is fraught with all sorts of potential problems. In old houses, flat and plumb walls are rare. (For that matter, you don't know what to expect in a new house, either.) The supporting wall directly behind the drywall or plaster can be anything from natural stone masonry to rough-hewn studs. Worse yet, the materials and their dimensions could be different on each side of the firebox. Your best bet is to prepare for all possibilities by having each of these anchoring methods on hand in your tool box.

Simple Colonial Mantel

Once their most pressing need for basic shelter had been satisfied, early Colonial settlers began to decorate their modest homes. And since the fireplace was such a central component of their lives, its appearance took on considerable significance.

The fieldstone hearths typically found in 17th-century homes were sometimes crudely framed with plain pine boards in a vain effort to dress up the wall's rough character and provide a surface for nails and hooks for hanging pots and utensils. Later, it was simple to add a narrow shelf supported by brackets to the plain frame—a perch for the family's prized pewter. This was sufficient for a working kitchen that hosted an array of domestic activities.

But as daily life became more comfortable, the settlers developed more attractive approaches to decorating the fireplace. As demonstrated in this project, simple vertical paneling with a molded shelf makes for a streamlined and understated fireplace surround. Chair rail and base molding on the adjacent walls, both stopped short of the fireplace opening, add linear contrast and help define this Colonial hearth. While this project was built into a corner hearth, the design would work well on most fireplaces.

You can buy off-the-shelf paneling and wood moldings for this project. Milling your own will result in a more refined look. Combining shopmade and store-bought components offers a practical compromise.

Simple Colonial Mantel

THE COLONIAL MANTEL has two distinct elements—the paneled wall (you may need to add furring strips first) and the built-up mantel shelf (which is supported by a plywood box). The base molding finishes off the project with style.

FRONT VIEW

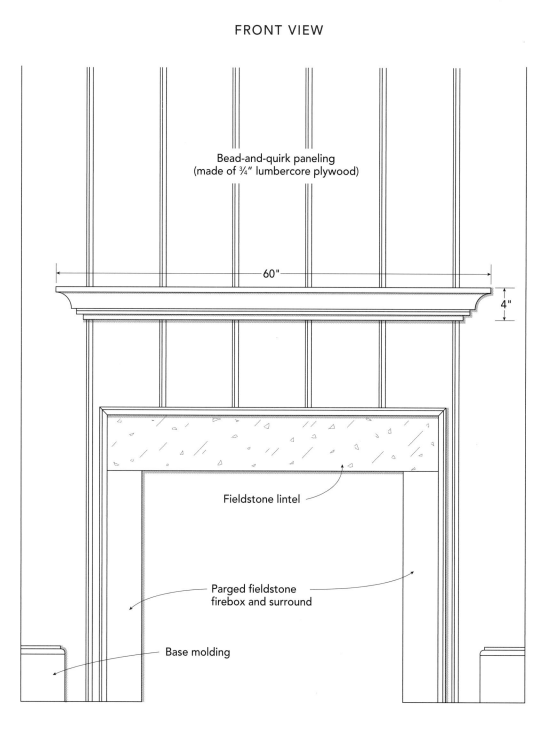

Bead-and-quirk paneling
(made of ¾" lumbercore plywood)

60"

4"

Fieldstone lintel

Parged fieldstone
firebox and surround

Base molding

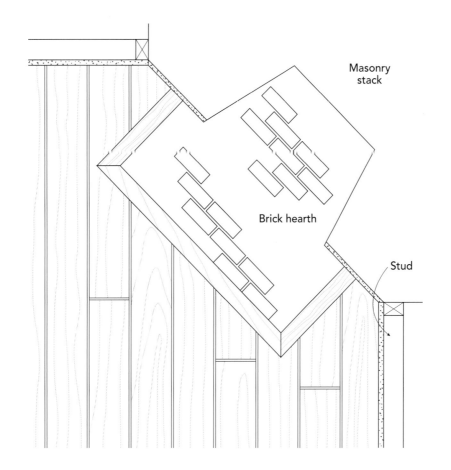

Masonry
stack

Brick hearth

Stud

BUILDING THE MANTEL STEP-BY-STEP

This seemingly complex shelf is actually a combination of relatively simple moldings.

This project has two distinct phases—making and installing the paneling, and then adding the mantel, which is built up of several layers of molding. The chair rail and base molding are options you can handle in different ways according to your room conditions and still keep the same basic look.

MAKING THE VERTICAL PANELING

While the earliest wall paneling was composed of square-edged boards butted together, an overlapping edge joint prevents air infiltration. The shiplap joint may be the easiest overlapping joint to make, but I chose to use a tongue-and-groove joint here and I added a bead-and-quirk detail along each seam for decorative effect. (See sidebar "Designing Vertical Panels" on p. 26.)

CHOOSING MATERIALS

For the colonists, the problem of wood shrinkage—and the resulting gaps—was unavoidable. The material available was not well dried to begin with, and the shrinkage was further complicated by the heat generated from the fire and through the chimney stack. Back then, having gaps in your wall paneling wasn't a big deal. But standards today call for straight lines, level and plumb construction, and clean results. So to avoid wood movement problems altogether, I used lumbercore plywood instead of solid pine for the paneling. (See p. 10 in chapter 1 for more on plywood.) A particular advantage of using lumbercore plywood for paneling is that each individual plywood "board" was dead straight when ripped from the sheet on the table saw. So when you install the paneling, you won't have to muscle each board into position on the wall.

For the moldings, I used mainly off-the-shelf pine purchased from a local home center. However, the three bead profiles (one around the foundation opening, one under the mantel shelf, and one on the chair rail) were not available, so I made these pieces myself on the router table.

MANTEL SHELF CROSS-SECTION

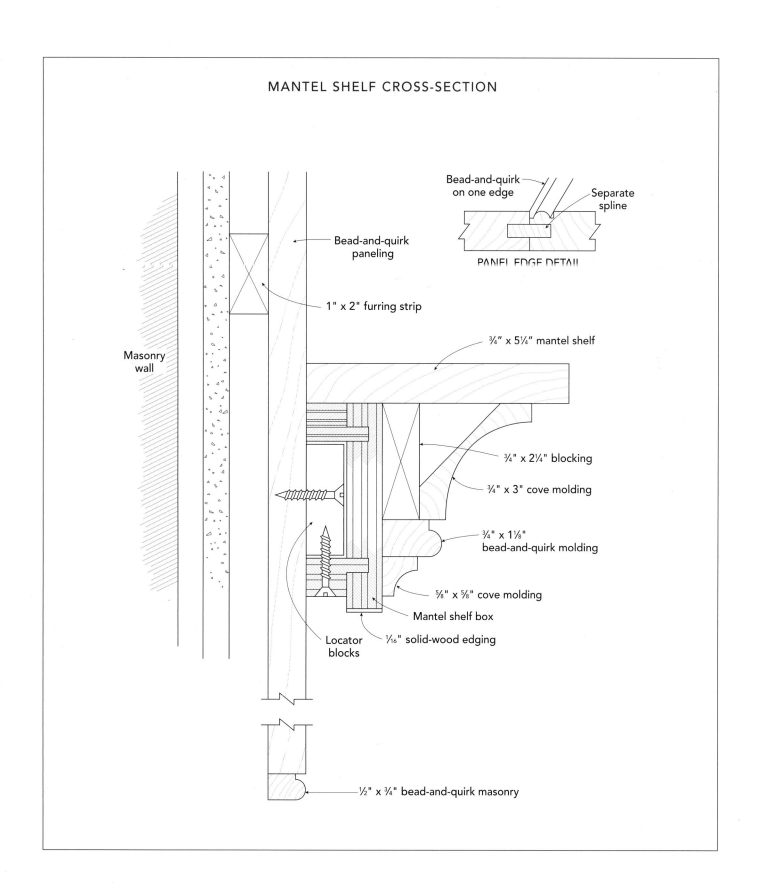

Masonry wall

Bead-and-quirk paneling

1" x 2" furring strip

Bead-and-quirk on one edge

Separate spline

PANEL EDGE DETAIL

¾" x 5¼" mantel shelf

¾" x 2¼" blocking

¾" x 3" cove molding

¾" x 1⅛" bead-and-quirk molding

⅝" x ⅝" cove molding

Mantel shelf box

1/16" solid-wood edging

Locator blocks

½" x ¾" bead-and-quirk masonry

Masking tape holds the spline in the groove while the glue sets.

To form the panel edge joint, cut a groove in each edge, rout a bead on one edge, and fit a separate spline in the groove. Test one spline before gluing them into all the boards.

Ripping the paneling to width

1. Decide on the width of the panels by doing a horizontal layout on a story stick or mocking up a segment of the wall with sample panels. The end pieces can be narrower than the rest but should be equal to one another.
2. Cut all the panel pieces to width on the table saw.

Routing the panel grooves

Instead of overlapping the edges of the paneling or cutting a tongue-and-groove joint, you can cut a groove in both edges and use a loose spline. That way there's only one setup involved. You can cut the grooves on the table saw, but supporting such long boards will be difficult. Instead, clamp them to a workbench and cut the grooves with a router, using a ¼-in. slotting bit.
1. Clamp the panel board flat on a bench. You can also use a set of sawhorses with a length of 2-by stock for longer support.
2. Set the router up so the cut is centered in the panel stock's thickness.
3. Rout the grooves in both edges of each board.

Routing the bead

With a bead-and-quirk–cutting router bit, the ball-bearing guide controls the lateral depth of cut, so you only have to set the vertical depth.
1. Clamp the board vertically to the bench or in the vise, with the edge several inches above the bench surface. This ensures that the router will clear the bench and allow more comfortable access to the workpiece.
2. Rout the bead. Be sure to hold the router level and guide it firmly along the edge of the board in order to cut a smooth and fully formed bead.
3. Go over the routed beads with 120-grit sandpaper to remove any imperfections or fuzz.

Cutting the splines

You may be able to use ¼-in. plywood for your splines. Exterior-grade fir plywood and Masonite® are usually a true ¼ in. thick, though many plywoods are slightly thinner and won't work. You can also cut the splines from solid wood.

1. Rip the splines to width on the table saw.

2. Glue the splines to the beaded edge of each board. Apply glue to the groove, and slip the spline into the groove. Hold the spline in place if necessary with masking tape.

3. Before gluing in all the splines, fit a pair of boards together to test the fit.

4. Prime the paneling with a heavy coat of water-based primer. This raises the grain and any remaining fuzz. When the primer dries, sand with 120-grit sandpaper, then apply a second light coat of primer.

INSTALLING THE PANELING

Plumbing the wall

Whatever the condition of the existing wall, the completed paneling must be straight and plumb. The best method of establishing the proper foundation for the paneling is to use horizontal furring strips (1 by 2 or 1 by 3). The furring strips may need to be shimmed out in places to create a flat and plumb nailing surface.

1. Drop a plumb line from the ceiling to the floor about 2 in. from the wall. If there is any point along the wall that measures more than 2 in., attach a shim to the wall to make up the difference. Use construction adhesive and a couple finish nails to secure the shims. Repeat this process every 2 ft. along the wall.

2. Mark the furring strip locations. For an 8-ft.-high wall, plan on five strips and mark out the wall accordingly. Place one strip 3 in. from the ceiling, another 3 in. from the floor, and one about every 2 ft. in between.

3. Attach the furring strips directly to the masonry wall with masonry screws.

Hanging the paneling

1. Start with the second board in from the end—the end boards will get scribed to fit last. Use a level to mark a plumb line where the edge of this board will lie on the wall.

I primed, filled, and sanded the paneling before installation.

This detail shows the chair rail terminating neatly on the wall paneling.

DESIGNING VERTICAL PANELS

Colonial house carpenters employed a variety of joints and edge treatments for vertical paneling. All these profiles—bevels, beads, and grooves—were executed with wooden bench planes and molding planes. These simple tools made a linear cut that produced distinctive shapes by repeatedly passing the plane along the board's edge until the profile was complete. It was a quick, direct method, executed with the available woodworking technology of the day.

Shown here are three examples of joints used in vertical paneling during the 17th and 18th centuries. There were many regional design variations, as well as local variations based on basic realities such as the tools owned by individual craftsmen. This selection can be used as inspiration for designs of your own. Many edge treatments can be re-created by mixing router bit profiles to come up with identical or similar patterns. Or you can seek out some old molding planes for a more authentic reproduction.

What about panel width? Two hundred years ago, random-width boards were typically used for paneling. In an early-American interior, the boards might measure anywhere from 8 in. to 18 in. wide. Availability dictated the dimensions of the boards, and personal aesthetics determined their arrangement.

Unless you live in an antique house, I would mill all the boards to a uniform width. On this project, I initially measured the wall and divided it evenly into 6-in.-wide boards. Then I laid out the arrangement on a story pole, a stick that has all the critical dimensions inscribed on its surface. Although I liked the heavy and close repetition of vertical lines, I thought it would be a little too busy and modern-looking. So I reduced the number of boards and increased the width to 9 3/4 in. This change gave the wall a cleaner appearance that would allow the mantel shelf to stand out better.

PANEL JOINT VARIATIONS

Attach furring strips securely to the masonry wall to form a straight and plumb foundation for the paneling.

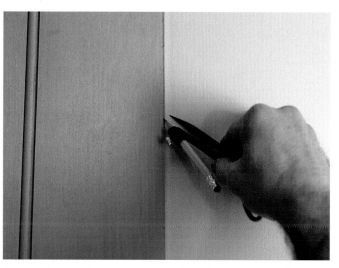

Scribe the end boards to the wall with a compass, then saw or plane the edge to fit.

The paneling is plumbed and then secured to the furring strips with narrow-crown staples.

2. Cut all the boards to length and nail the first one in place. I attached the boards with 1¼-in.-long narrow-crown staples nailed through the face. These small fasteners go into the material easily and leave a small countersunk entry hole.

3. Scribe the two end boards. Hold the boards in place and use a compass to scribe the contour of the end wall. Then cut the edge of the end boards with a saber saw.

MAKING THE MANTEL SHELF

The mantel shelf is the focal point of this fireplace. The shelf should be elegant, yet establish a stylistic relationship with the plain paneled wall. I decided to keep the moldings simple but give the shelf some height and visual weight. The structure of the shelf is basically an open-backed box. The visible exterior of the box is covered with moldings, and then the finished mantel is mounted onto blocks attached to the paneling.

Building the shelf box

I wanted a stable foundation for the shelf moldings, so I built the shelf box out of ¾-in. plywood (see the drawing on p. 23).

The plywood shelf box provides a solid foundation for the decorative shelf moldings.

This miter jig cradles the large cove molding for a clean and precise cut with a handsaw.

Using a pin nailer gives a secure joint with minimal damage to the workpiece.

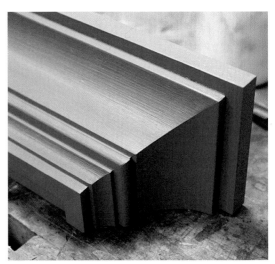

Prime, fill, and sand the assembled shelf. When I've achieved a neat, clean job, I apply the finish coat of paint.

1. Cut the parts for the shelf box to size.
2. If using plywood, edge-band the bottom edge of the front of the box with either a thin strip of solid wood or iron-on veneer tape.
3. Join the parts of the box together. Although it was a little more trouble than butt joints and screws, I used a rabbet-and-dado joint instead.
4. Add the spacer block that supports the large cove.

Adding the mantel moldings

The mantel is made up of four layers: the mantel shelf, a large cove, a ½-in. bead, and a ⅝-in. cove. When assembling mitered moldings, it's a good idea to "walk" the molding around the project, starting on the left and working around to the right.
1. Cut and fit the bead molding against the bottom of the spacer block. Use a finish nailer or pin nailer to attach the moldings to the

shelf box. Both of these pneumatic nailers use fasteners so small that they seldom split moldings and leave a tiny entry hole that's easy to disguise later. You can also use finish nails and a nail set, however.

2. Repeat the process with the small cove molding, and nail it in place under the bead.

3. Cut the mantel shelf and nail it to the plywood box and spacer block.

4. To cut the cove molding, I held it steady with the help of a miter jig.

5. After all the moldings are nailed to the mantel box, fill any gaps, dings, and nail holes with a water-based filler, sand everything smooth with 180-grit sandpaper, and then prime and paint the shelf.

Installing the mantel shelf

1. Decide on the proper height for the mantel shelf and mark a line on the wall.

2. Screw mounting blocks to the paneling to support the mantel shelf. I cut these blocks to fit exactly into the back of the mantel shelf box.

3. Screw the mantel shelf onto the blocks from underneath, and countersink the screws.

Adding bead-and-quirk molding

I dressed up the panel opening around the fireplace with quirk-and-bead molding. This also neatly concealed the gap between the paneling and the masonry. Cut the molding to fit with miters at the inside corners, and nail it in place.

THE CHAIR RAIL AND BASE MOLDING

Two hundred years ago, chair rail and base molding served the same purpose they do today: to decorate the room and protect the walls from unintentional damage caused by the movement of furniture and people. The chair rail is an assembly of simple moldings, one set on top of another. There is a nice progression, from a 3-in.-wide beaded foundation molding to a slender intermediate molding with a cove cut along each edge, and crowned with a slender half-round. The base molding is composed of a plain board with a base cap molding on top.

The installation of locator blocks simplifies hanging the finished shelf onto the paneled wall.

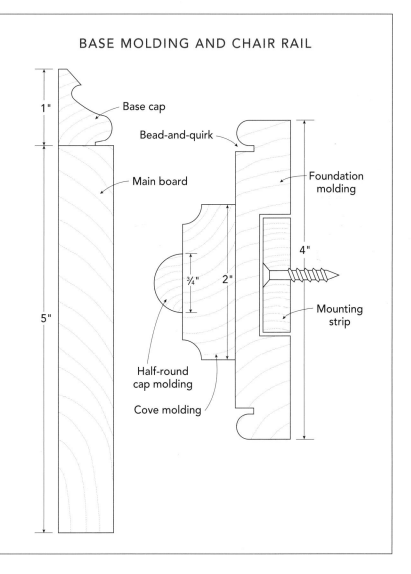

BASE MOLDING AND CHAIR RAIL

Base cap

Bead-and-quirk

Main board

Foundation molding

Half-round cap molding

Cove molding

Mounting strip

1"

5"

3/4"

2"

4"

Shown is a cutaway arrangement of the chair rail moldings. First the mounting strip is secured to the wall, then the decorative moldings are attached over the mounting strip.

This detail of the base molding is scribed to the floor and neatly returned to the paneling.

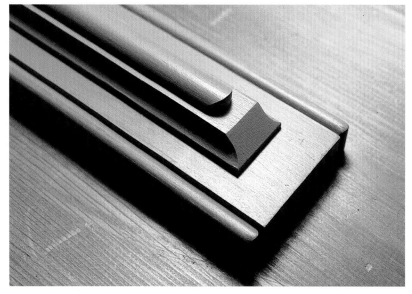

The chair rail molding terminates neatly with a combination of returns, angled cuts, and roundovers.

Installing the chair rail

1. Attach a level mounting strip that will support the chair rail moldings. Secure the mounting strip to the wall with large screws (toggle or hollow-wall fasteners could be used instead).

2. Make the chair rail foundation piece. Plain 1 by 4 or 1 by 6 stock will work fine, depending on the width you want. Rip the stock to width, cut the channel in the back face to fit over the mounting strip, and then rout the bead-and-quirk profiles on each edge. Nail the foundation piece to the wall.

3. Add the cove molding. Use a spacer block to position the cove molding consistently at the center of the foundation molding.

4. Nail on the cap molding.

Terminating the chair rail

On the mantel wall, the chair rail turns the corner from a plastered wall onto the paneling and then stops short of the mantel shelf by about 6 in., terminating with a sculptured treatment that looks finished, not abruptly stopped or cut off.

1. For the foundation molding, miter the end and attach a return, cut from the same material and placed over the exposed miter cut.

2. Bevel the intermediate molding at an attractive angle—35 degrees looks good.

3. Cut the half-round square, then round the ends over to look as if a return has been added.

A bead-and-quirk molding nicely trims the fire-place paneling.

When coping an inside miter joint, start with a plain 45-degree cut.

Installing the base molding

Like the chair rail, the base molding turns the corner and continues onto the paneled wall, stopping short of the mantel.

1. Cut the base molding to length and nail it to the wall. Stop the base molding about 4 in. from the edge of the firebox opening.

2. Cut the base cap molding and nail it in place.

3. Inside miters are susceptable to opening up over time, so I use a coped joint here, even on a 135-degree angled wall like this. To cut a coped joint, let the first piece (either side) butt into the corner. Cut the second piece at a 45-degree angle. Then cut along the exposed edge of the miter with a coping saw or jeweler's saw so this piece overlaps the first piece.

4. End the base cap with return pieces.

5. Fill any remaining visible nail holes, and give the mantel a final coat of paint.

The outline of the profile is then coped with a jeweler's saw.

GEORGIAN MANTEL

During the late 1700s Samuel MacIntyre was a master carver and furniture maker whose work adorned the most beautiful homes in the Boston area. His carving was characterized by luxuriant flowers and vines combined with linear bands of short vertical flutes, and his signature device was the wheat sheaf. His carvings were unique because he successfully combined classical architectural details with lifelike natural elements.

MacIntyre's masterpiece was the design, construction, and decoration of the Pierce-Nichols house on Federal Street in Salem, Massachusetts. He began the commission in 1782, at the height of the American Georgian period, but didn't complete the interiors until 1801. In that year, he executed a superb mantel in the more refined Adams style for the drawing room, which was being prepared especially for the wedding of Sally Pierce, the daughter of a wealthy shipper, to George Nichols. The Pierce-Nichols drawing room mantel was the inspiration for this project.

Although this is clearly the most ornate mantel in the book, it's not the most difficult to build. The main parts are made of paint-grade birch plywood, edged with solid wood, then embellished with applied moldings and ornaments. While the moldings are made of wood, the "carved" ornaments are actually made of composite material cast in a mold. These cast architectural products are easy to use, finely detailed, and relatively inexpensive.

Georgian Mantel

THE CAREFULLY DETAILED ORNAMENTS AND MOLDINGS on this mantel appear carved, but they're made from composite materials (ordered from a catalog) and applied to a simple plywood foundation. The result is a mantel of classic proportions and refined decoration.

FRONT VIEW

10"

Upper pilaster

58"

Lower pilaster

7"

8"
plinth

2⅛"

PLAN VIEW

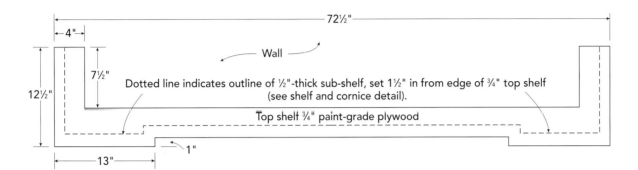

72½"

4"

Wall

7½"

Dotted line indicates outline of ½"-thick sub-shelf, set 1½" in from edge of ¾" top shelf (see shelf and cornice detail).

12½"

Top shelf ¾" paint-grade plywood

13"

1"

REED DETAIL

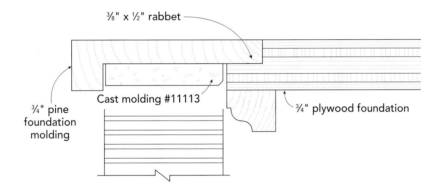

⅜" x ½" rabbet

¾" pine foundation molding

Cast molding #11113

¾" plywood foundation

BUILDING THE MANTEL STEP-BY-STEP

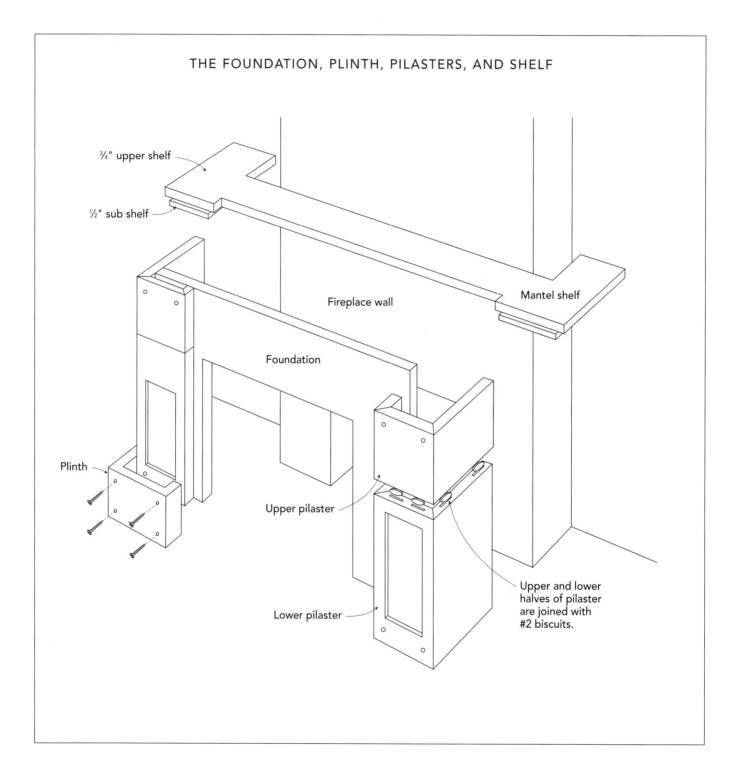

THE FOUNDATION, PLINTH, PILASTERS, AND SHELF

¾" upper shelf

½" sub shelf

Fireplace wall

Mantel shelf

Foundation

Plinth

Upper pilaster

Lower pilaster

Upper and lower halves of pilaster are joined with #2 biscuits.

This mantel is built on a typical foundation, made of ¾-in. plywood, which provides a stable surface for the various moldings and decorative elements. The columns and plinth blocks that flank the opening are separate L-shaped constructions that wrap around the foundation and return to the wall. Crowning the assembled mantel is a double shelf. Onto these flat surfaces a variety of wood and cast moldings and carvings are applied.

My plan was to construct, assemble, and paint as much of the mantel as possible in the shop, then transport the completed foundation, pilasters, plinths, and shelves to the site for final assembly and installation.

THE FOUNDATION

This foundation is a three-part U-shaped construction that frames the firebox opening and is screwed to furring strips on the fireplace wall. It is the stage onto which all the various parts of the project are carefully brought together.

Assembling the plywood foundation boards

The horizontal foundation board rests on the two vertical boards, and the parts are joined with biscuits.

1. Cut the foundation parts to size.

2. Cut biscuit slots in the top ends of the two vertical foundation boards and corresponding slots in the bottom edge of the horizontal board.

3. Glue and clamp the foundation parts together. Ensure that the parts are square, and secure the assembly by screwing a cleat across the back of the verticals at the bottom.

4. Rout a ½-in.-wide by ⅜-in.-deep rabbet along the back edge of the foundation opening (see the bottom drawing on p. 35). A bearing-guided rabbeting or slotting bit works best. But if you don't have one that cuts the full ½-in. width, clamp a straightedge to the back of the foundation and use a straight cutting bit.

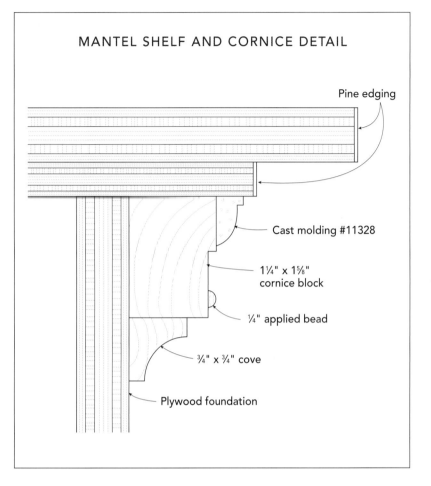

MANTEL SHELF AND CORNICE DETAIL

Pine edging

Cast molding #11328

1¼" x 1⅝" cornice block

¼" applied bead

¾" x ¾" cove

Plywood foundation

Milling the foundation molding

The foundation molding creates a solid bed for the short horizontal reeds that frame the fireplace opening. It also protects the fragile reeded molding from unintentional collisions with fireplace tools and firewood. The foundation molding is easily cut from solid 1-by pine on the table saw in a two-part operation using a combination blade.

1. Cut the flat foundation molding stock to size. I used a 1 by 4 ripped to 3 in. wide.

2. Cut the short shoulder ½ in. from the edge and ⅜ in. deep.

3. Reset the fence for a ⅜-in. cut and raise the blade about 1½ in. high. With the workpiece on edge and the waste side of the molding on the free side of the blade, cut partway up to

Glue the L-shaped foundation molding into the rabbet in the back of the foundation boards. The assembly forms a protective trough for the compo reeding that frames the fireplace opening.

Fit the mitered reed molding around the foundation (see the sidebar on p. 42 for instructions on working with composite moldings).

the shoulder cut. Then raise the blade and make a second pass to complete the cut.

4. Cut the foundation molding to fit around the foundation assembly, mitering the inside corners.

5. Cut biscuit slots in the miters to reinforce these joints, then glue and clamp the foundation molding to the foundation assembly.

This detail shows the compo ornament applied to the pilaster.

6. Miter the reeded molding and glue it onto the foundation. Also add the mitered trim piece around the outside of the reeded molding.

THE PILASTERS AND PLINTHS

The pilasters are made in two parts. The upper pilaster section is made of ¾-in. birch plywood, mitered at the corners, while the lower section has a recessed center panel that receives a cast swag element. The sides, or returns, of the pilasters extend to the wall, perpendicular to the fireplace opening.

Making the upper pilasters

1. Cut the upper pilaster parts to size. For the short inside return pieces, start with a wider piece that will be easier to handle when you're cutting the edge miter. Miter the edges of the

PLINTH AND PILASTER ASSEMBLY DETAILS

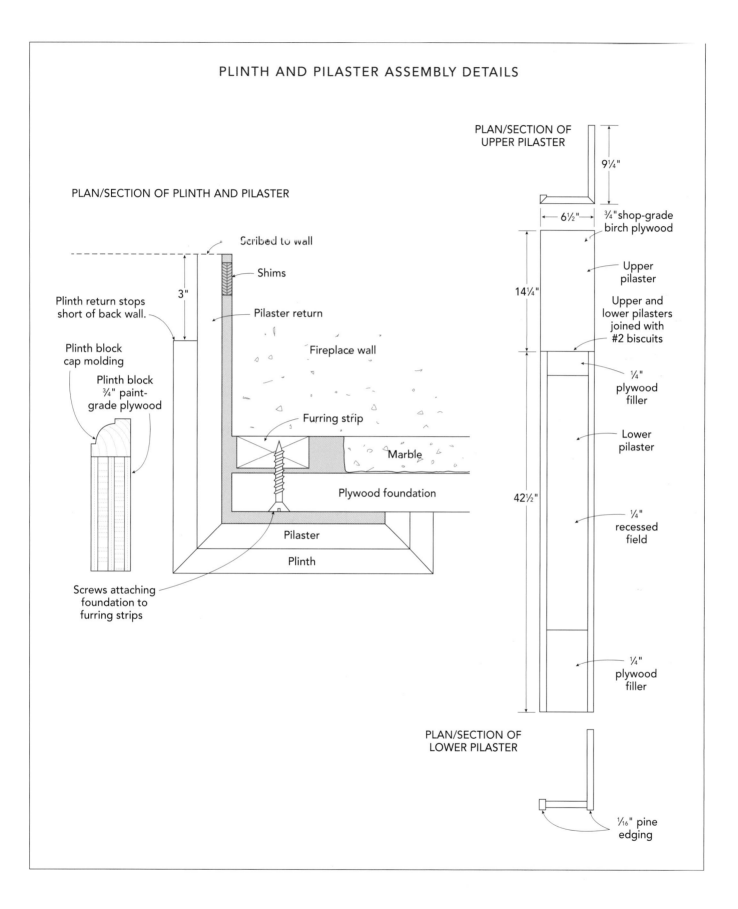

PLAN/SECTION OF
UPPER PILASTER

9¼"

PLAN/SECTION OF PLINTH AND PILASTER

Scribed to wall

Shims

3"

Plinth return stops
short of back wall.

Pilaster return

Plinth block
cap molding

Fireplace wall

Plinth block
¾" paint-
grade plywood

Furring strip

Marble

Plywood foundation

Pilaster

Plinth

Screws attaching
foundation to
furring strips

6½"

¾"shop-grade
birch plywood

Upper
pilaster

14¼"

Upper and
lower pilasters
joined with
#2 biscuits

¼"
plywood
filler

Lower
pilaster

42½"

¼"
recessed
field

¼"
plywood
filler

PLAN/SECTION OF
LOWER PILASTER

¹⁄₁₆" pine
edging

Cover the edges of the lower pilasters with thin solid-pine edging. Masking tape works better than any clamp.

The mitered cuts of the plinth blocks are glued and taped together.

(only 14¼ in.), I simply aligned the corners by hand, rubbing the joints together, and secured them with masking tape (see the bottom photo at left, where the same technique is used for assembling the plinths).

Making the lower pilasters

1. Cut the lower pilaster parts to size. When ripping the face pieces, make sure the assembled lower pilaster will be the same width as the upper pilaster.

2. Glue edgebanding onto the front edges of the side pieces. I used solid pine and yellow glue, and clamped the edgebanding on with masking tape. Iron-on edgebanding material is a good choice if you have it available.

3. Cut biscuit slots in the parts so the face piece of each assembly will be recessed ¼ in. from the edge of the side pieces.

4. Glue the three parts together.

5. Cut the ¼-in.-thick filler stock to fit in the top and bottom of the recess, then glue it in place to form the framed recess for the cast swag.

Joining the upper and lower pilasters

1. Cut a couple biscuit slots in the face and long return pieces of the upper and lower pilasters where they join together.

2. Assemble the two sections to complete the pilasters. Don't worry about getting a tight seam here, because it gets covered by the frieze molding. But do make sure the assembly is straight by holding a long straightedge against the face and side.

Making the plinths

The plinth blocks wrap around the pilasters. They're made of ¾-in. plywood, mitered at the corners, then capped with a solid wood molding. I decided to stop the plinth returns 3 in. short of the back wall, creating an interesting detail and allowing the back wall base molding to tie into the pilaster returns. The plinths are constructed identical to the upper pilaster sections, with mitered corners and rubbed glue

parts. (See p. 69 for my method of sawing miters.)

2. Rip the inside return pieces to 1 in. wide.

3. Assemble the upper pilasters. You could cut slots in the miters and use splines to align these miters, but because the pieces are short

The mantel shelf is made from three wood profiles that can be purchased or shaped on a router table. The bead-and-quirk molding is a commercially available composite.

joints. Be sure to allow an extra $\frac{1}{16}$ in. on the front piece so the assembled plinths will fit easily over the pilasters. I covered the exposed plywood edges with solid pine molding.

BUILDING THE MANTEL SHELF

The mantel shelf is constructed in two pieces. The lower shelf is ½-in. plywood and overhangs the bead-and-quirk molding by $\frac{3}{16}$ in. The upper shelf is ¾ in. and overhangs the lower shelf by 1½ in. (See the drawing on p. 36 and the Plan View on p. 35).

1. Cut the mantel shelf parts to size.

2. Lay out and cut the 1-in.-deep notch in the front edge of both parts. I made the short 1-in. cuts with a handsaw, then made the long stopped cuts on the table saw. This involves laying the piece down on the spinning blade at the start of the notch, then stopping the cut at

The plinth blocks are capped with a pine molding. To ensure a tight fit and a clean joint, I glued the molding pieces together before attaching them to the plinths.

WORKING WITH COMPOSITE MATERIALS

This mantel is covered with bead-and-quirk molding, garlands and swags, margents, and vases. Imagine the work and skill necessary to carve all this decoration in wood by hand—not to mention the expense.

Instead I chose precast ornaments, called composition ornamentation or "compo" in the trade. These are decorative elements that resemble wood carvings but are actually molded from wood fibers and resins. The material can be molded to duplicate almost any carving in intricate detail. And the completed castings can be stained to resemble wood, or painted. Unless you closely examine the cast ornaments, you'd be hard-pressed to distinguish them from those carved of wood.

I selected the compo ornaments for this mantel from The Decorators Supply Corporation in Chicago, although composite architectural moldings and ornaments are available from many suppliers.

The ornaments arrived carefully packed in a padded box and taped onto cardboard backing sheets. When I removed them for inspection, they were hard and brittle. Some of the linear moldings were not very straight or flat. Initially, I thought this would be a problem, but everything turned out fine.

No special equipment was needed to install the ornaments or moldings. In fact the products come with the glue already on the back. All the process required was a hot plate (or other source of heat) and a large shallow metal pan to hold water, and a piece of muslin attached to a frame. (Don't use a metal screen because the composite products can melt right onto the screen.)

To attach composite elements, you simply apply steam by resting them on the muslin screen, which softens them and activates the hide glue on the back. The softening allows the moldings to be bent around curves or straightened if necessary.

The composition ornaments closely resemble real wood carvings, though they are more fragile.

The products arrive expertly packed to avoid any damage in shipping as well as when you start to work with them.

While they're still warm, position the moldings and ornaments onto the wood surface, then press them firmly against the wood surface.

That's it. No nails, special adhesives, or complicated procedures were required. When the compo ornaments cooled, they were fixed permanently and ready for finishing.

I cut the linear compo moldings on the miter saw with a fine-tooth blade, dropping the spinning blade onto the material slowly and with extra care. Occasionally the brittle molding would splinter, but the resulting damage was easy to repair later. And any discrepancy was easily fixed during installation, when the material was soft and very pliable.

I didn't follow any strict order of application. But I found the larger ornaments, like the vases and margents (the floral pieces on the pilasters), to be more forgiving and easier to place or move around than the smaller linear moldings. So I started with those, and as my confidence and skill developed, my assistant and I progressed to the smaller, more tedious linear moldings.

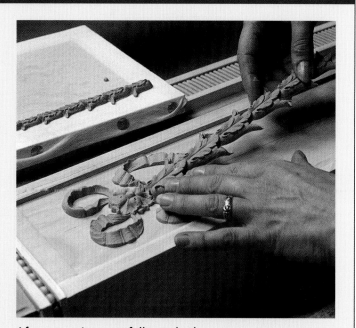

After steaming, carefully apply the compo ornaments to the wood surface. A few moments of pressure bonds the ornaments in place. No nails or special fasteners are needed.

To apply the compo ornaments, you first carefully steam them over a pan of water. The steaming makes the material pliable and also activates the hide glue on the back.

A miter saw can be used to cut or miter the cast moldings. To avoid splintered or broken edges, lower the blade slowly into the material.

The edges of the plywood mantel shelves are covered with thin solid-pine strips. Use masking tape to secure the strips while the glue dries.

The plinth returns stop 3 in. short of the back wall, allowing the wall base molding to tie into the pilaster returns.

the end of the notch. Raise the blade just high enough to cut through the material, and be cautious when lowering the workpiece onto the spinning blade. An alternative approach is to make the long cut on the bandsaw, then trim the cut with a straightedge and flush-trimming router bit.

3. If needed for your installation conditions, cut the end return pieces and attach them to the main shelves with biscuit joints.

4. Apply edgebanding to all the edges of both shelf parts.

5. Trim and sand the edgebanding, then screw the lower shelf to the upper one.

Preparing the cornice moldings

The mantel shelf is supported by a four-part cornice molding assembly. Three of the profiles are wood, but the bead and quirk is a compo molding. The three wood shapes can be purchased, or you can shape your own from solid stock on the router table. Note that the main cornice block (see the drawing on p. 37) is rabbetted to create a step just below the bead-and-quirk profile. Prepare enough of each profile for the mantel. (Castings supplied by The Decorators Supply Corporation, Chicago, Illinois.)

PREASSEMBLY

When the foundation, pilasters, and plinths were ready, I assembled the main mantel parts. My objective was to end up with all the moldings attached to the mantel in advance of installation, but with the pilasters removable to make delivery and installation a little easier. You can, of course, put the mantel together once and install it as a single unit. Even then you'll want to install the plinths on site so they can be scribed to the floor if necessary, and also leave off the cornice cove molding so you can hide the installation screws behind it.

Attaching the pilasters and plinths

1. Lay the foundation flat over a pair of sawhorses. Position the pilasters and screw them to the foundation, placing the screws where they will be hidden behind the plinths, the frieze, and the cornice moldings.

2. Check the fit of the plinths but don't attach them yet.

THE FRIEZE BAND

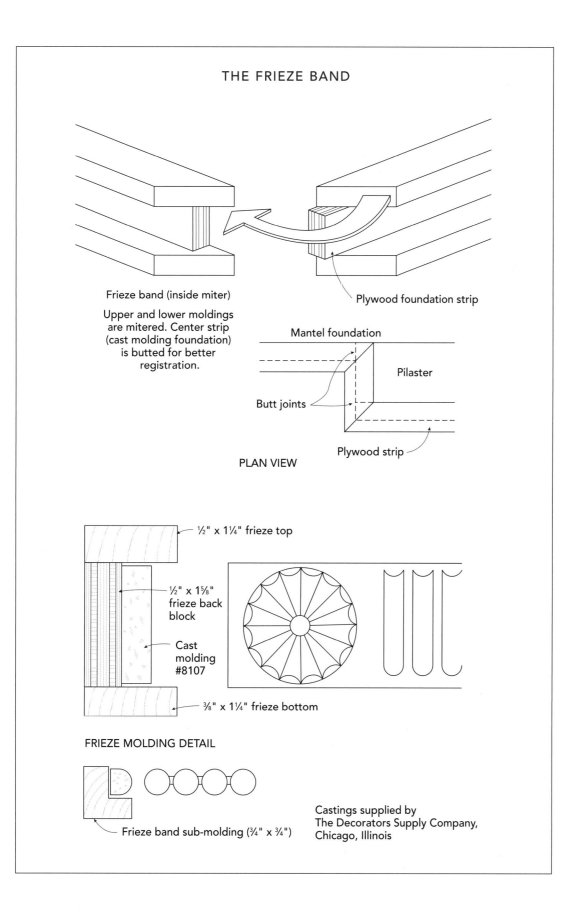

Frieze band (inside miter)

Upper and lower moldings are mitered. Center strip (cast molding foundation) is butted for better registration.

Plywood foundation strip

Mantel foundation

Pilaster

Butt joints

Plywood strip

PLAN VIEW

½" x 1¼" frieze top

½" x 1⅝" frieze back block

Cast molding #8107

⅜" x 1¼" frieze bottom

FRIEZE MOLDING DETAIL

Frieze band sub-molding (¾" x ¾")

Castings supplied by The Decorators Supply Company, Chicago, Illinois

Miter the main cornice block, then attach it with screws to the plywood foundation.

Attach the small cornice moldings with a brad nailer or pin nailer.

Applying the frieze components

The frieze is composed of two strips of compo molding, each supported by blocking. The main band fits in a U-shaped trough, while the smaller band rests in an L-shaped strip of solid wood. Each element is cut and fit individually around the pilasters, and nailed to the foundation.

1. Cut sufficient lengths of each blocking material as shown in the drawing on p. 45.
2. Cut and nail together the U-shaped trough assembly that supports the main frieze molding. Because I would be disassembling the pilasters from the mantel, I used an interlocking miter joint at the corners, as shown in the drawing on p. 45. It takes more work, but the joints really went back together perfectly at installation. An alternative approach is to assemble the entire run of the U-trough, and miter through the whole assembly at the corners. Attach the frieze foundation with nails.
3. Cut a rabbet in lengths of solid stock to form the L-shaped submolding, then cut and fit this piece around the mantel.
4. Cut and apply the two compo moldings to the frieze band.

Applying the cornice moldings

1. Cut a shallow rabbet in the face of the main cornice block, then fit it around the top of the mantel.
2. Apply the other cornice moldings in the same way, working around the mantel from one end to the other. I left the cornice cove molding off the face of the pilasters in order to hide my installation screws.

PAINTING THE MANTEL

From its inception, this mantel was going to be painted. So the visual dissonance created by the use of different materials, each chosen for a particular reason, was not important. Once painted, the completed mantel is a unified piece of architectural woodwork.

1. After sanding with 220-grit sandpaper, apply two coats of water-based primer, sanding between coats.
2. Fill any remaining dings, gaps, and surface blemishes with a water-based wood filler.
3. After sanding the primed and patched mantel, wipe a thin bead of caulk along all the main joints. This step softens the corners, tones down any glaring seams, and gives the

Tip: To enhance the working and drying properties of your paint, try adding a commercial product like "Floetrol," which improves the flow and leveling of the paint. Providing you avoid drips, the finished surface levels to a nice flat sheen.

The plywood, solid pine, and compo ornaments present a visual jumble, but a high-quality paint job will disguise the differences in color and surface texture.

finished project the look of a carefully preserved antique mantel.

4. Paint the mantel with a premium-grade latex enamel.

INSTALLING THE MANTEL

Your fireplace will have its own particular conditions for installing the mantel, but the procedure followed here is fairly typical.

1. Attach furring strips to the wall, shimming them out as necessary. This is commonly required to create a level and plumb surface, or to build out even with the masonry around the fireplace opening.

2. Set the foundation in place evenly on the firebox opening, and screw it to the furring strips.

3. Position the pilasters in place and carefully align the molding elements. Scribe the outside

After screwing furring strips to the fireplace wall, set the foundation in place. Make sure it is even left-to-right over the firebox opening, then screw the foundation to the furring strips.

Position the pilasters, complete with moldings and ornaments, in place against the foundation. With the bulk of the work done in the shop, the installation took only minutes.

returns if necessary. Screw the pilasters to the foundation.

4. Place the mantel tops in position, and scribe the back edges if necessary. Then screw down into the pilasters with trim-head screws to secure the tops.

5. Nail on the cove molding at the top of the pilasters.

6. Slip the plinths over the pilasters and scribe them to the floor if necessary, then screw the plinths to the pilasters.

7. Fill the exposed screw and nail holes with a latex wood filler or acrylic caulk and touch up with paint after the filler dries.

Both sections of the mantel shelf are screwed into the pilasters.

The last piece of molding is pin-nailed into place, covering the pilaster screws and completing the cornice.

Screw the plinth blocks into the pilasters, and then fill, sand, and paint over the small countersunk holes.

ALTERNATIVE CORNICE MOLDING DESIGNS

Faced with designing a cornice (defined as the uppermost of three primary divisions of the classical entablature), many woodworkers simply look over their collection of router bits, pick the most elaborate profiles, and pile one on top of another. The result is often a visually dense conglomeration of random profiles that adds nothing but weight to the top of the design.

Sometimes it's the space between the moldings or the areas left plain and square that make a great cornice design. I'm not an architect, nor do I pretend to be an expert on classical architecture, but I always try to have a little fun with my cornice designs. I'll just play with a particular combination of moldings, changing their order or size, until everything clicks.

One device I use is to extend vertical and horizontal lines from the edges of the molding profiles, as shown in the drawing below. The grid formed by these lines should yield a number of squares or nearly square rectangles, which indicates a smooth transition from one shape to the next and an overall balanced projection. The drawing at top right is a good example. Long and narrow rectangles, as shown in the other three drawings, indicate too little or (in these cases) too much projection.

This quick technique doesn't follow any strict rules of classical proportion or design and won't necessarily result in a masterpiece cornice. But it does provide a way to establish some basic proportions that might prove helpful when designing one. The most foolproof method would be to study and copy a respected work of art or architecture.

SIMPLE FEDERAL MANTEL

This mantel is typical of those found in many rural farmhouses in the early 19th century. Almost always made of wood and painted, the style was taken directly from classical architecture and imitated the design of basic shelter: columns supporting a beam and roof. The simple moldings and joinery indicate that it could have been built by a local carpenter instead of by a furniture joiner. But its simplicity doesn't diminish its appeal in any way. The mantel's flat relief and plain treatment perfectly frame the Federal-period hearth opening and provide a focal point for the display of family possessions and a backdrop for social gatherings and important events. The mantel's design shows elegant proportion, restraint, and balance. And the simple moldings cast bold shadows that highlight its timeless appeal.

The federal mantel is structurally straightforward and can easily be built in a weekend. Three boards joined together with biscuits form the foundation, which is fastened to the wall. Plinth blocks (doubled-up 1-by stock) support the plain vertical pilasters, which support the horizontal architrave. Add a few moldings and the mantel shelf, and you're ready to paint.

Simple Federal Mantel

PROVING THAT SIMPLICITY DOESN'T PRECLUDE ELEGANCE, this mantel design is anchored by ideal proportions and perfect symmetry with the brick firebox opening it adorns. Built with readily available materials and
moldings, it's easy to build as well.

FRONT VIEW

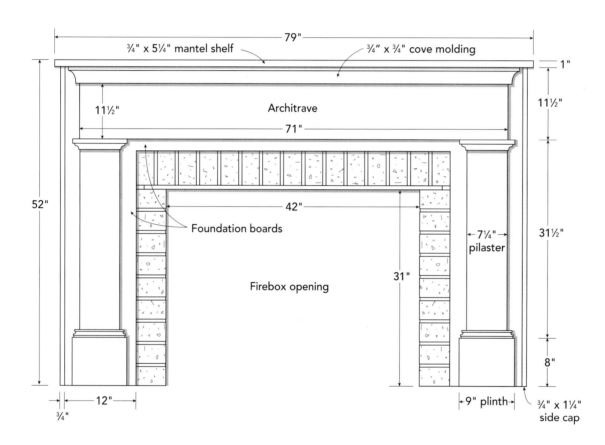

¾" x 5¼" mantel shelf   ¾" x ¾" cove molding

79"

1"

11½"

Architrave

11½"

71"

52"

42"

Foundation boards

7¼"
pilaster

31½"

31"

Firebox opening

8"

12"

9" plinth

¾"

¾" x 1¼"
side cap

BUILDING THE MANTEL STEP-BY-STEP

Begin by preassembling the foundation board and laminating the plinth blocks, you can move directly to installation. I chose to preassemble some of the molding elements as well.

THE FOUNDATION BOARD

The foundation board is the backdrop of the mantel. It provides a flat surface for the mantel proper, and bridges any gaps or irregularities between the masonry and the adjacent wall surface, while exposing only the neatest brickwork. The mantel foundation was designed with the lintel section fitting between the columns. That way the mantel parts would overlap the foundation joints, making the whole construction stronger.

1. Cut the two columns and lintel that will form the foundation. The firebox opening in this project is 32 in. high by 42 in. across, and an even course of bricks is left exposed around the sides and top. Using a 14-in.-wide lintel (horizontal section) and 10½-in.-wide columns (vertical sections) produced the balanced proportions that form the basis for the mantel's design. You should adjust these dimensions based on the size of your firebox opening.

CHOOSING MATERIALS

During the 19th century, pine was abundant and readily available, and carpenters used it for most interior trim, including fireplace mantels. So a meticulous reproduction would require large, wide boards of clear pine. However, the use of solid pine for this project would present problems (besides price) for the modern woodworker that 19th-century carpenters weren't concerned with.

At that time houses weren't insulated, so warm and cold air passed through the structure freely. In a particular room, it wasn't unusual to experience surprising differences in temperature. With a fire blazing in the hearth, the warmest spot in the room would have been a seat in front of it, while other areas of the same room might be as much as 15° colder. These conditions surely played havoc with human comfort but spared furnishings and interior woodwork from drastic changes in temperature and humidity. In a modern ultra-insulated home, wood is subjected to extremes of temperature and relative humidity created by efficient central heating and air-conditioning. The use of wide, solid boards and true period construction methods in a modern home would probably cause unsightly checking and splitting. Miters would likely open up, and flat sections would cup.

A better approach for today's woodworker would be to construct this mantel using lumbercore plywood instead of solid wood. I used ¾-in. lumbercore plywood for everything except the plinth blocks and the moldings. (See chapter 1, pp. 9–12, for a detailed discussion of materials.)

Join the foundation boards with a couple of biscuit slots.

Tip: You'd think pieces of molding stock at a lumber store are all identical. But if there are pieces from different batches, there could be slight differences, which will result in miters that don't line up perfectly. To avoid this, I try to cut all my mitered pieces from the same length of stock so there's no doubt that the profile is the same on all the pieces.

2. Lay out and cut biscuit joints to connect the lintel to the columns—three or four #2 biscuits should do the job.

3. Glue up the foundation assembly, making sure the columns are square to the lintel. When the assembly is dry, remove the clamps; but before moving it, attach two support battens across the front. The battens reinforce the joints, maintain the dimensions of the foundation opening, and keep it flat during installation.

THE PLINTH, PILASTERS, AND ARCHITRAVE

Laminating the plinth blocks

The plinth blocks at the base of the pilasters are made with two pieces of ¾-in.-thick solid pine laminated face-to-face. The net 1½-in. thickness is needed to support the pilaster and the plinth molding. You could use a chunk of 2-by stock, but the approach here resulted in a more stable block, plus it made good use of scrap material I had on hand.

1. Cut the plinth block pieces slightly oversize.

2. Saw or rout two grooves into the back face of each piece, about 1½ in. from the edges.

3. Fit a spline into each groove, and glue the mating surfaces together.

Cutting the parts to size

1. Arrange the main mantel parts (pilasters, architrave, and plinths) on the foundation.

2. Center the parts and cut them to length.

3. Cut biscuit joints to align the top of the pilasters to the architrave.

4. Cut the plinth blocks to size. (Depending on the condition of the hearth, you may want to leave the plinth blocks a little long so they can be scribed to the hearth at installation.)

Selecting the moldings

I purchased stock moldings from the local building supplier. The simple profiles I needed were readily available, in quantity. By choosing available profiles instead of choosing special-

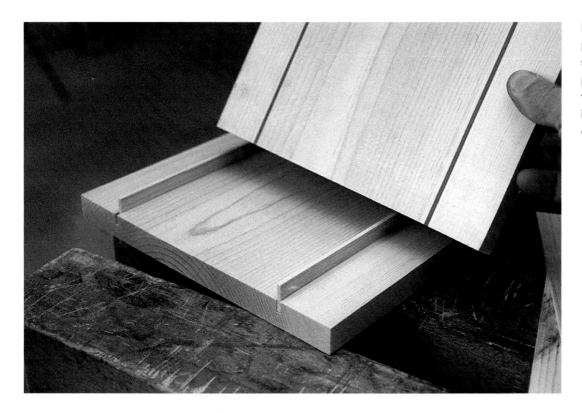

Laminating two pieces yields a more stable plinth block. A pair of splines keeps the pieces from sliding around when clamping up.

order profiles, I could pick through the inventory and select the straightest and cleanest material.

There were three distinct profiles I needed: a large and simple cove for the cornice molding, an ogee with fillet for the torus molding (at the base of the pilaster), and a large ogee with quirk (space or reveal) for the capital molding. These last two moldings are both sold typically as "base cap" profiles.

Priming the parts

To achieve an attractive painted surface, the wood components must be carefully prepared. This involves filling any holes and dents and repairing cracks. I do some of this after installation, but it's easier to do a first go-over now. Also, on this mantel I primed the moldings before cutting and fitting them to the mantel.
1. Fill any holes, dents, split seams, tearout, or cracks in your material with a water-based wood filler. On lumbercore plywood, I usually apply filler on the exposed edges, paying par-

The flexible blade on a good-quality putty knife will fill any voids in the material and not further mar the surface.

The finger joints, visible on the edges of the lumbercore, should be filled and sanded before you attach the parts to the mantel.

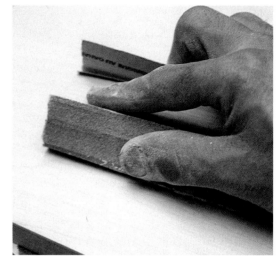

Use a large half-sheet sander or a sanding block to level any primed surfaces. Break square edges slightly but don't round them over too much.

All moldings should be filled, primed, and sanded for the best appearance.

Tip: If a water-based filler dries up, you can easily rehydrate it with a little tap water. You can even change the consistency if you prefer a thinner filler.

ticular attention to the finger joints where the solid material was spliced.

2. When the filler is dry, I use a medium-grit (120 to 150) sandpaper to remove any excess and then level the surface.

3. Clean off the filled and sanded boards with a tack rag, then apply a water-based paint primer. For a fluid coating that lays down nicely, I thinned the primer about 20 percent.

It can be applied with either a brush or a roller. The primer fills and levels the wood and raises the grain slightly.

4. When the primer dries, look for any flaws that might have been missed the first time around, and fill them. Apply a second thinned coat of primer, and when dry sand again with 150-grit to 180-grit paper. Now the surface is ready for paint.

INSTALLING THE MANTEL

Anchoring the foundation

Unless your walls are flat and plumb and you can determine the location of the studs behind, attach furring strips to the wall first, then attach the foundation to the strips. That way the principal method of attachment, no matter what you choose, will eventually be hidden by the mantel parts. In this case the brick masonry surrounding the opening was ½ in. higher than the surrounding plaster wall. In order to make up this difference and give myself a tiny margin, I cut my furring strips to ⅜-in. thickness.

1. Attach furring strips to the wall. The furring strips can be secured with lead anchors, masonry screws, or cut nails.

2. Position the foundation against the wall, and center it on the opening.

3. Check the foundation for plumb and level, then screw it to the furring strips with #8 wood screws. Locate the fasteners so they'll be covered over by the other mantel parts later.

Building up the mantel

With the foundation securely in place, you can apply the next layer of mantel parts. Working from the bottom up may seem more logical, but I worked from the top down and scribed the plinth blocks to the floor last.

1. Attach the architrave to the foundation with 1¼-in. screws. Make sure the top edge is even

Furring strips, shimmed plumb as needed and attached to the wall surface, provide good solid support for the foundation. Use the appropriate fastener based on the wall material.

Position the braced foundation against the furring strips. Make sure it's plumb and leveled, then screw it to the strips with #8 by 1½-in. wood screws.

This detail shows the capital molding that caps the pilasters.

ARCHITRAVE-PILASTER JOINT

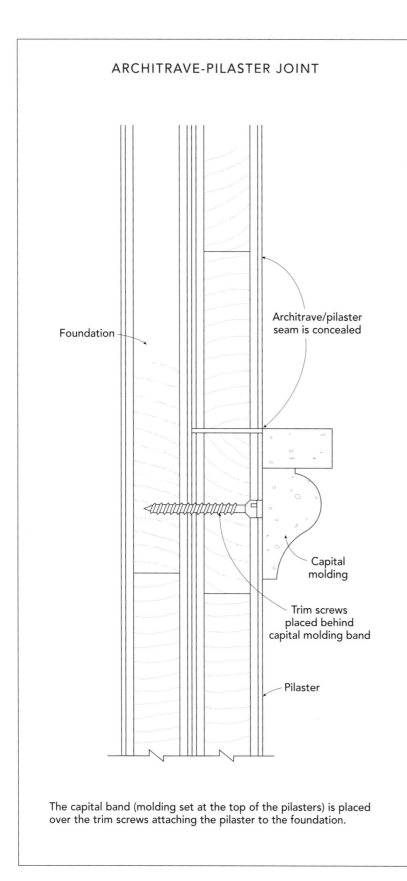

Foundation

Architrave/pilaster
seam is concealed

Capital
molding

Trim screws
placed behind
capital molding band

Pilaster

The capital band (molding set at the top of the pilasters) is placed
over the trim screws attaching the pilaster to the foundation.

**With the architrave in place, set the pilasters,
using biscuits for alignment and added strength.**

with the foundation board and that the spaces
at the ends are equal.

2. Position the pilasters under the architrave,
and add the biscuits and glue to reinforce the
joint. Secure the pilasters to the foundation
with 1¼-in. screws. Locate the screws at the
bottom and top of the pilasters, where they'll
be covered over with the capital and torus
moldings.

3. Fit the plinth blocks. Once the pilasters are
in place, measure the remaining space for the
plinth blocks. On both sides of this mantel
there was a small discrepancy between the
wood floor and the slightly raised brick of the
hearth. So I scribed the ends of the plinths to
fit, made the cut with a jigsaw, and attached
them to the foundation with countersunk
trim screws.

PLINTH

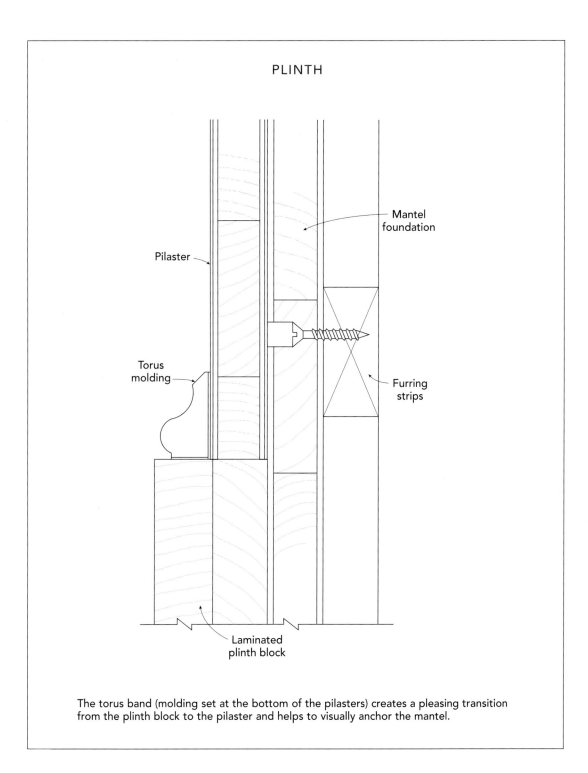

Mantel foundation

Pilaster

Torus molding

Furring strips

Laminated plinth block

The torus band (molding set at the bottom of the pilasters) creates a pleasing transition from the plinth block to the pilaster and helps to visually anchor the mantel.

Blocking for the cove molding

In order to provide a stable bed for the cornice molding, I made up some blocks to be placed along the top edge of the frieze and under the mantel shelf. The 45-degree face of these blocks supported the cornice molding at a consistent angle and ensured that the miters would line up properly. To support the small return sections of the cornice, I added a small piece of wood to the back of the angled blocking.

1. Saw the cove blocking from a piece of 2-by stock. Make sure the angle of the blocking

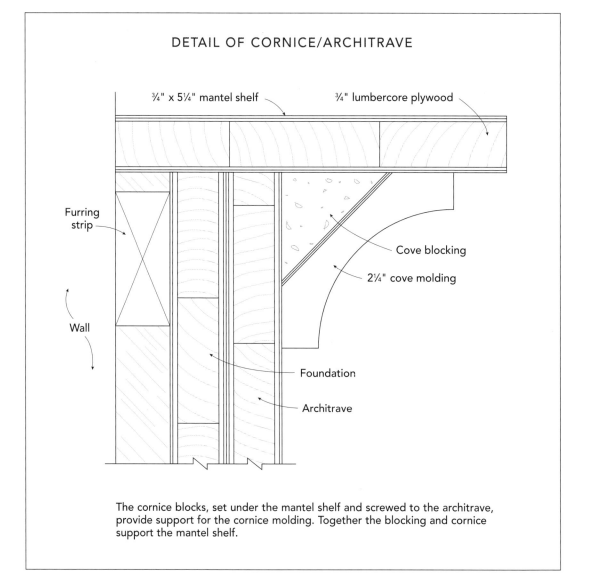

DETAIL OF CORNICE/ARCHITRAVE

¾" x 5¼" mantel shelf

¾" lumbercore plywood

Furring strip

Cove blocking

2¼" cove molding

Wall

Foundation

Architrave

The cornice blocks, set under the mantel shelf and screwed to the architrave, provide support for the cornice molding. Together the blocking and cornice support the mantel shelf.

A small block is glued to the angled cove blocking. This supports the cornice molding return piece.

Screw angled cornice blocks along the top edge of the architrave.

PREASSEMBLED MOLDING BANDS

On any project, moldings attract my attention. I always look to see whether the profile matches up and wraps around the corner cleanly. And of course, I like to see tight miters. If you're laying down the molding as you go, this is sometimes difficult to achieve. To make the job easier, I often build my bands first and then attach them to the mantel.

By mitering, gluing, and nailing the bands together first, you can coax tight joints at the corners, allow them to dry, and then fill and sand them. All of this critical work is a lot easier if you can freely adjust the molding band. In addition, once the band is dry, it will flex slightly and conform to its position on the mantel—while the miter remains tight. And the constructed band will stay in place with fewer nails than if it were laid up one piece at a time.

I cut the sections on a miter saw to within $\frac{1}{32}$ in., then I plane them to fit with a low-angle block plane. When I'm satisfied with the fit, I glue the miters and nail them together with a pin nailer. I use a fixed block as a guide to assemble the pieces.

A preassembled band of molding can be gently coaxed into place—while the miter remains tight.

This detail shows the plinth with the torus molding.

After setting the legs of the compass to the widest gap between the mantel shelf and the wall, drag the compass along the length of the shelf. Here the mantel shelf is still oversize, so the scribed amount is a full inch larger than the widest gap.

face matches the angle of the cove molding you're using.

2. Attach the cove blocking through predrilled holes with trim-head screws.

THE MOLDINGS AND MANTEL SHELF

The conventional approach to installing moldings is to work your way around the mantel from one side to the other, fitting one piece to the next. (For an alternate approach, see "Pre-assembled Molding Bands" on p. 61.)

The mantel shelf

In the 18th and 19th centuries, woodwork was attached to the studs, then the walls were plastered, with the woodwork acting as a gauge or stop. The finish coat of plaster was then brought up to the woodwork. This method produced an interesting junction where the woodwork and plaster met that was soft and easy on the eye. But today's woodworkers and finish carpenters scribe their work to conform to the walls.

1. Set a compass to the width of the widest gap between the straight edge of the shelf and the wall.

2. With the pin leg of the compass resting against the wall and the pencil leg on the mantel shelf, pull the compass along the wall and shelf. This will result in a pencil line on the shelf that will mimic the wall surface.

3. Cut along the pencil line, then use a plane or rasp for final fitting.

The cove molding

I cut the cove molding on a miter saw outfitted with a special support carriage to hold the molding at the correct angle.

1. Cut the cove molding to fit.

2. Nail the cove to the cove blocks and mantel shelf with finish nails. Add some glue to the miters to help hold the joints closed.

3. When cutting the short return miter, make the 45-degree cut on a longer piece, then make the square cut to release the return from the longer stock.

The capital and torus moldings

1. Cut and fit these moldings around the pilasters.

2. Use a finish nailer for the long pieces and a pin nailer (or just glue) for the short returns.

3. Cut the side cap molding, and nail it to the edge of the foundation board. If necessary, scribe it to fit cleanly against the wall.

PAINTING THE MANTEL

Final preparations

With the mantel primed, sanded, and installed, there might be small gaps where the various sections of the mantel meet. Although they don't appear unsightly now, these gaps will stand out later and will work against a clean and unified appearance when the mantel is painted.

1. Fill any exposed screw or nail holes with putty.

2. Use a high-quality water-based caulk (Phenoseal® brand takes paint beautifully) in an applicator gun to apply a small continuous bead anywhere there is a gap. Within minutes of applying the caulk, wipe away any excess with a damp rag.

Applying finish coats

I used a water-based latex paint for the final coating of the mantel. For a project like this, I don't think oil-based paint offers any great advantages. I wanted a smooth surface with just a hint of brush marks that would imitate the finish on period woodwork.

The secret to a good job is to take your time, so I decided to apply the paint in several light coats. A thin coat levels nicely and dries more quickly and completely than a single heavy coat. I thinned out the paint about 20 percent and used a good-quality 2-in. synthetic brush. I started on the edges, then did the inside corners, and finished up with the large flat areas. Wait until each coat is thoroughly dry before proceeding with the next coat. The whole mantel required three coats of paint and a couple of 15-minute touchup sessions.

Nail on the capital molding with a pin nailer. Don't try to nail the miter or the wood may split.

The finish coat of paint should be applied in several thin layers. A thin coat of paint will level out nicely and dry quickly.

ARTS AND CRAFTS MANTEL

Many fireplace mantels done in the Arts and Crafts style are relatively simple designs. They often exhibit architectural details from other areas of the house. So there is a repetition of architectural motifs, such as molding combinations or brackets. This thematic repetition creates a familiarity from room to room that is comforting and reassuring. In the best examples, the Arts and Crafts style fosters a feeling of well being and calm that carries you through the house, barely aware that you've left one space and entered another.

Overall, the look of the Arts and Crafts style is one of restraint and subtlety. The style doesn't rely on carvings, turnings, or book-matched veneered panels to catch the viewer's eye. The construction is direct and honest. Its success relies on the contrast between the gentle curves that mimic nature and the stout, straight architectural members that suggest stability, strength, and permanence.

This mantel is an abbreviated design that doesn't call for much material or take a lot of time to build. It doesn't touch the floor or reach the ceiling. It wraps itself around the middle of the firebox wall like a belt, accenting the fireplace without overpowering the rest of the modest woodwork in the room. And in this case, the brevity of the mantel beautifully showcases the rich mottled green Arts and Crafts tile.

Arts and Crafts Mantel

BUILDERS FROM THE ARTS AND CRAFTS ERA frequently employed decorative tile around a fireplace. While a traditional fireplace design would conceal some of the tile, this mantel allows the tile (or stone or brick) to perform front and center and instead forms a crowning adornment entirely above the firebox. Clean lines and honest joinery capture the essence of an Arts and Crafts interior.

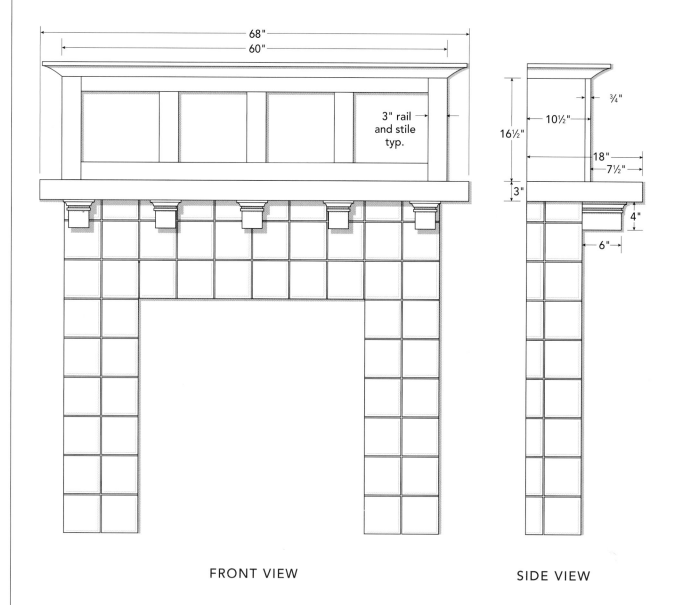

FRONT VIEW

SIDE VIEW

BUILDING THE MANTEL STEP-BY-STEP

There are four distinct parts to this mantel: shelf, overmantel, brackets, and cornice molding. Each is relatively easy to build. The four elements are then installed one at a time over the fireplace opening.

BUILDING THE SHELF

Mission fireplace mantels typically had a thick solid shelf. I opted for a plywood construction technique because it would be more stable. It also gave me an easy way of anchoring the shelf to the wall. My design calls for four layers of ¾-in.-thick oak-veneered plywood. The core of the shelf is two layers of ¾-in. shop-grade plywood, glued and screwed together. The outer oak layers are then wrapped around the core, creating a recess at the back of the shelf. This recess fits over a cleat, attached securely to the wall (see the drawing on p. 68).

Laminating the core
1. Cut the two pieces of the shelf core to size.
2. Screw and glue these pieces together.

Mitering the shelf edges
I wanted a seamless shelf, so I utilized a method of cutting the layers beneath the oak veneer to 45 degrees, while leaving the very top of the oak veneer intact. The drawing on p. 69 shows how to set up your saw to cut the plywood. One advantage to this method is that you can't "overcut" the mitered edge, but you can recut the edge for a better fit without cutting into the top veneer and altering the surface dimensions of the workpiece.
1. Cut all the outer shelf parts to size.
2. Using the fence setup shown in the drawing on p. 69, miter the front edge and both ends on the shelf top and bottom.
3. Miter-cut all four edges of the shelf front. When mitering the short ends, use a miter gauge to guide the stock through the cut.

Tip: When laminating two pieces of plywood, check whether the pieces are bowed. Glue them together with the concave faces inward, and the laminated assembly should be much straighter than the individual pieces had been.

CHOOSING THE MATERIAL

Oak was the principal wood used for both furniture and architectural fittings of the Arts and Crafts style. Because it gave the work an honest and vigorous look reminiscent of the romantic medieval style that inspired the English Arts and Crafts movement, oak was adopted here in America by builders and craftsmen as the wood of choice.

During the early 20th century, quartersawn stock was used almost exclusively. Although this material is available today, expect to pay a premium for it. Despite its cost, though, I wouldn't use anything else. Quartersawn oak displays an open-grain surface without the distracting grain patterns of the cheaper and more common flatsawn variety. I used ¾-in. plywood for the mantel shelf assembly and ¼-in. plywood for the overmantel panels. Staining the project allows the plywoods to blend well with the solid wood components.

SIDE CROSS-SECTION

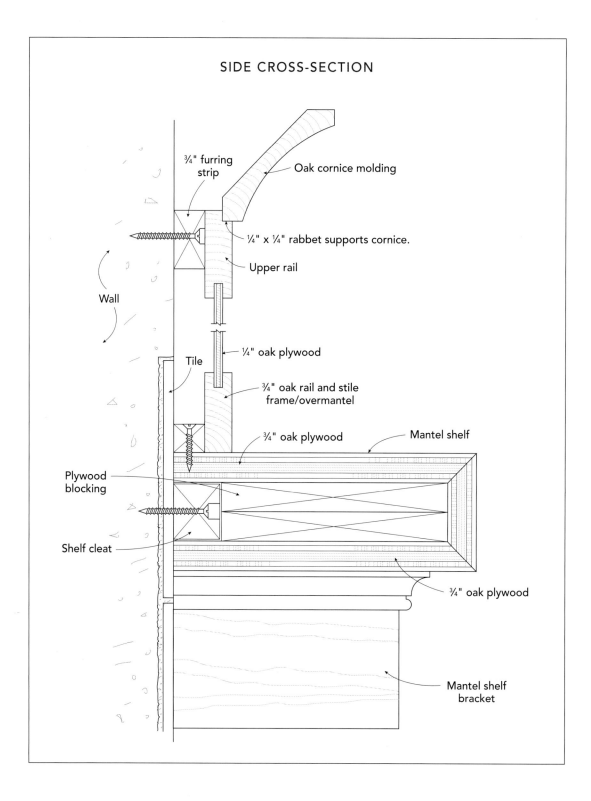

¾" furring strip

Oak cornice molding

¼" x ¼" rabbet supports cornice.

Upper rail

Wall

Tile

¼" oak plywood

¾" oak rail and stile frame/overmantel

¾" oak plywood

Mantel shelf

Plywood blocking

Shelf cleat

¾" oak plywood

Mantel shelf bracket

Cutting Miters in Plywood

SETTING UP THE TABLE SAW

1. Attach a plywood fence to your table saw fence and mark the thickness of the ¾-in. oak plywood onto the plywood fence.
2. Set the table saw blade to 45 degrees and position the plywood fence to receive the table saw blade, at an angle, just below the scribed line.
3. With the saw running, raise the blade, cutting into the fence slightly.
4. Run some test pieces and check the fit of the mitered edges, one to the other.

Caution: You must use a push block to complete this miter cut and remove the waste piece from under the blade. Simply push a scrap piece of plywood (6x12 will do comfortably) against the back edge of the workpiece as if you were going to miter the edge of the push block. Continue cutting into the push block until the waste piece falls free. Then carefully slide the push block back from the blade while holding it tightly against the fence and table.

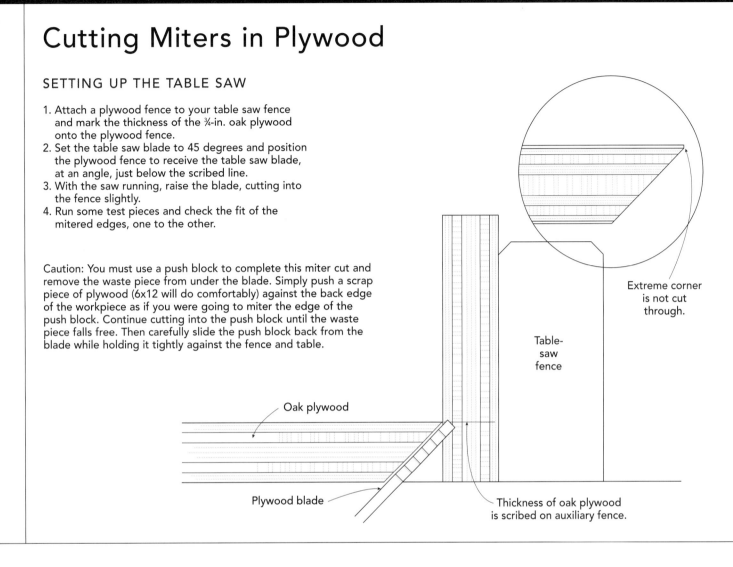

Extreme corner is not cut through.

Table-saw fence

Oak plywood

Plywood blade

Thickness of oak plywood is scribed on auxiliary fence.

4. Miter the top, bottom, and front edges of the two end pieces.

Assembling the shelf

For the best results, all the main shelf parts should be assembled at the same time. (The returns are assembled separately.) That way any discrepancies, gaps, or other problems can be corrected. It's worth clamping the parts together dry to make sure the fit is right.

1. With the inside surface of the top layer facing up, position the plywood core on top of it so the edge of the core is even with the inside edge of the miters.

2. Screw through the core into the shelf top with #8 by 2-in. wood screws.

Before assembling the shelf, test-fit the miters around the shelf core.

Once the built-up core is carefully positioned, attach it to the top layer of the mantel shelf. Apply the front edge next.

This detail shows that the bracket appears to be a structural support, but it actually hangs from the mantel using a sliding dovetail joint.

RETURN ASSEMBLY

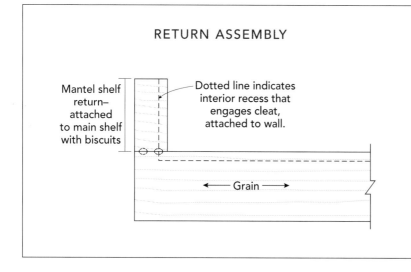

Mantel shelf return—attached to main shelf with biscuits

Dotted line indicates interior recess that engages cleat, attached to wall.

← Grain →

3. Attach the front of the shelf. I secured it with narrow crown staples, but finish nails would work as well.

4. Attach the bottom side of the shelf with glue and fasteners.

Closing the gaps

Naturally, there were some small gaps that needed attention. By completely assembling the shelf, I was able to "jiggle and tweak" the sections to close up the larger, more serious gaps. I closed the smaller openings by first forcing more glue into the space, then wiping any excess from the surface with a wet rag and using a burnisher to gently press the corner from both sides. When the gap is closed, secure the repair with either a clamp or a strip of masking tape.

Building the shelf returns

In my installation the firebox wall projected from the adjacent wall about 6 in. That meant the shelf would have to turn the corner and tie into the main wall for a clean-looking installation. (If your wall does not step out like this, omit the returns altogether.) Like the main shelf, the returns were four-layer laminations with a recess for the wall cleat. These returns

Because this fireplace has a built-up tile surround, shelf returns were needed to finish off the sides. These were made up separately, then attached to the main shelf with glue and biscuits. Later, the mitered end strips are glued on, spanning the joint between the main shelf and the returns.

The top side of the oak brackets are milled with a dovetail groove. The brackets are easily slipped over the dovetail cleats at installation.

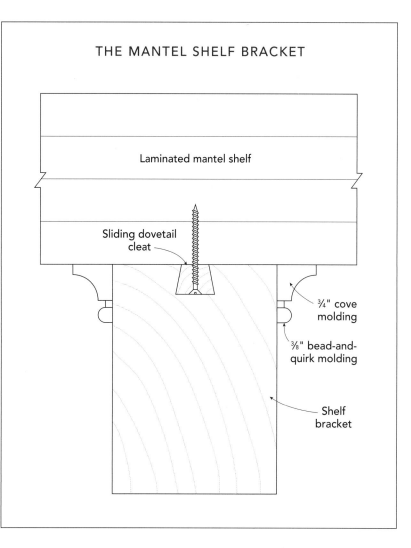

THE MANTEL SHELF BRACKET

Laminated mantel shelf

Sliding dovetail cleat

¾" cove molding

⅜" bead-and-quirk molding

Shelf bracket

were built separately, then attached to the main shelf with #2 biscuits. When the returns were permanently attached to the main shelf, the end pieces were glued and stapled on.

MAKING THE BRACKETS

The square-cut brackets appear to be jutting from within the wall and supporting the shelf. In fact, they are merely decorative and simply hang from the underside of the shelf. But they give the mantel a strong architectural feel. You could make the brackets from a single piece of 12/4 oak (as indicated in the drawing), but I built them up using ¾-in. stock instead. Try to achieve an even color and grain match in the end grain of these pieces. Each bracket was attached to the underside of the mantel shelf by means of a sliding dovetail cleat.

Laminating the bracket blanks

1. Cut the bracket blanks slightly oversize.
2. Arrange the blanks for best end grain match, then glue the four bracket blanks together.

3. Rip and crosscut the laminated brackets to final size.

Forming the dovetail cleat

1. Rout a ½-in. stopped groove into the top of each bracket. The groove should stop about an inch from the front edge of each bracket.
2. Re-rout the groove with a ¾-in. dovetail bit.
3. Use the same router bit to shape the edges of a ½-in. by ¾-in. strip so it fits snugly into the groove.
4. Round the end of the cleat with a disk sander or with files and a sanding block.

Adding trim to the brackets

Using the router table, I ran off two simple profiles for the bracket moldings—a cove and

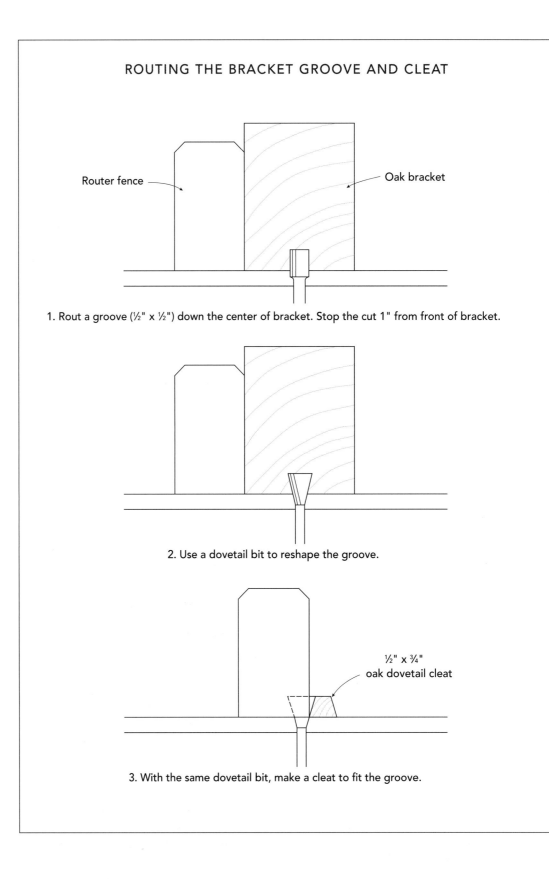

ROUTING THE BRACKET GROOVE AND CLEAT

Router fence

Oak bracket

1. Rout a groove (½" x ½") down the center of bracket. Stop the cut 1" from front of bracket.

2. Use a dovetail bit to reshape the groove.

½" x ¾"
oak dovetail cleat

3. With the same dovetail bit, make a cleat to fit the groove.

These details show the overmantel and cornice (left) and the mantel shelf (right).

The dovetail cleats that attach the brackets to the underside of the mantel shelf are milled with the same router bit used to mill the dovetail groove. After drilling a countersunk screw hole, bevel and round the end of each cleat to fit the recess.

a bead. Since the molding profiles were so small, I cut the necessary miters on a shop-made miter box using a 26-tpi saw. When all the cuts were made, I attached the moldings with yellow glue and a pin nailer.

BUILDING THE OVERMANTEL

This decorative section employs flat ¼-in. panels set into a rail-and-stile frame that wraps around the projecting wall. It bears no moldings or decoration itself, though it is capped off by the cornice. The simplicity of the overmantel contributes a look of modest substance to the mantel as a whole.

Making the frame

1. Cut the frame parts to size. Be sure to include extra length for stub tenons on the two long horizontal rails, and the three intermediate stiles.

Saw a ¼-in.-wide by ⅜-in.-deep groove in the frame stock pieces to receive the ¼-in. panels and the stub tenons.

The fence setting determines the shoulder cut. Then the waste can be removed by repeatedly passing the workpiece over the blade.

Set the height of the blade to cut the stub tenon shoulders by using a scrap of frame stock as a guide.

A shoulder plane is an efficient way to precisely fit the stub tenon to the groove.

2. Cut a ¼-in. groove ⅜ in. deep in the inside edges of the frame pieces, and both edges of the intermediate stiles.

3. Set the height of the table saw blade to cut the shoulders to form the stub tenons. Since the cheek of the tenon was so short, I simply ran the workpiece over the blade a few times to remove the waste and complete the tenon.

4. After all the joints are cut, test the fit. I used a shoulder plane to fine-tune the fit of the tenons into the shallow grooves.

Fitting the panels

1. Mark the locations of the three intermediate stiles on the inside edge of each rail to form four equal panel spaces.

2. Assemble the frame without glue, and clamp it together lightly.

3. Measure the openings for the panels, and adjust the stiles to make the openings equal if necessary.

4. Add ½ in. to each dimension to account for the grooves (¼ in. along each edge).

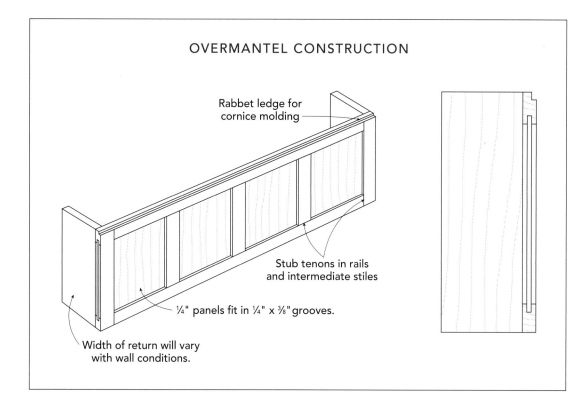

OVERMANTEL CONSTRUCTION

Rabbet ledge for cornice molding

Stub tenons in rails and intermediate stiles

¼" panels fit in ¼" x ⅜" grooves.

Width of return will vary with wall conditions.

Assembling the overmantel

1. Sand the panels, then glue the frame parts around the panels. Because the panels are plywood, they can be glued into the grooves. (Go lightly with the glue to avoid squeeze-out.) Make sure the assembly is square and flat.

2. Make the return pieces. My conditions required return panels at the sides of the overmantel that extended back to the wall. At the very least, you'll need a ¾-in. block to finish off the sides and conceal the furring strip space. Cut the return pieces to size and attach them to the back edge of the overmantel. I used biscuit joints.

3. Rout a ⅜-in. by ⅜-in. rabbet along the top outside edge of the overmantel, including around the side return. This provides a seat for the cornice molding. With the same setup, you can rout a rabbet on the inside back edge of the return; the reduced thickness makes scribing to fit much easier.

The return panels of the overmantel, made of ¾-in. oak plywood, are attached to the overmantel frame with biscuits.

After assembling the overmantel, rout a rabbet into the top edge to receive the cornice molding.

THE CORNICE MOLDING

The cornice crowns the overmantel and sits on a small rabbet cut into the top edge of the overmantel. I cut the molding on the table saw in order to achieve a good color and grain match and get the exact size I needed. For the tightest miter possible, I assembled the cornice as a complete unit before installation.

Milling your own cornice

Some lumberyards and home centers stock a selection of cove moldings in hard and soft wood ready for purchase, but I decided to make my own on the table saw. This is a basic table saw technique that is quite simple. It involves running the stock over the table saw blade at an angle, along a fence. Instead of cutting straight through, the blade will cut an elliptical arc-shaped plow into the underside of the workpiece. After each pass, the blade is raised a little until the desired depth and width of cut are obtained. The greater the angle of approach to the blade, the wider and flatter the cove. At 90 degrees to the blade, the elliptical cove flattens into a circular arc.

1. Mill the pieces for the corner stock to thickness and width. Draw the desired arc onto the end of a scrap piece.

2. Clamp a flat fence with a straight edge to the table saw as shown in the photo below. (I use a seat-of-the-pants approach to determining the fence angle: I raise the blade to the full height of the cut, lay the marked-out stock in front of the blade, then with my eye down close to the saw table and in front of the blade

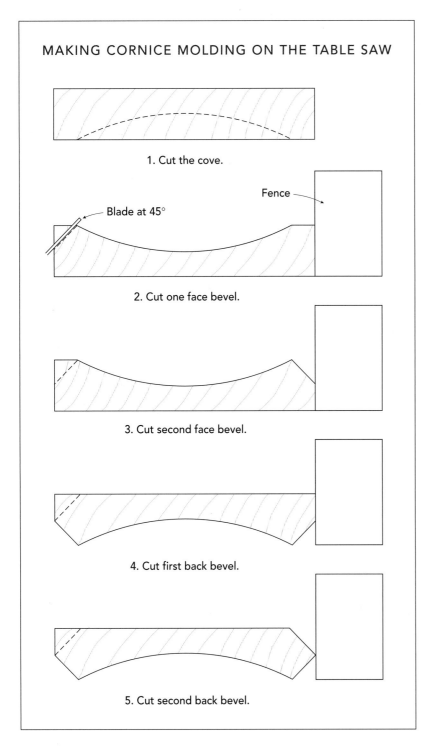

MAKING CORNICE MOLDING ON THE TABLE SAW

1. Cut the cove.

Fence

Blade at 45°

2. Cut one face bevel.

3. Cut second face bevel.

4. Cut first back bevel.

5. Cut second back bevel.

You can make the cove molding for the cornice on the table saw by passing the material over the rotating blade at an angle.

I adjust the angle of the workpiece until the cove on the stock corresponds to the arc of the blade.)

3. Set the blade height to about ⅛ in.

4. Run the cornice stock over the blade at a moderate speed.

5. Raise the blade about ⅛ in. and repeat the cut until the cornice cut is full-depth.

6. Make the four rip cuts at 45 degrees to complete the cornice profile.

Completing the cornice assembly

1. Sand and scrape the saw marks on the cornice moldings.

2. Cut the cornice pieces to fit around the overmantel (leave the return pieces a tad long).

3. Glue and nail the three pieces together, but don't attach the cornice to the overmantel yet.

After sawing the cornice molding, smooth the surface with scrapers and sandpaper.

RE-CREATING AN OLD FINISH

If you live in an old house, you might be lucky enough to rescue much of your original woodwork. But unless you decide to paint your trim (doors, casings, and paneling), you might be faced with the challenge of matching any new work to the old existing trim.

The first order is to determine a finishing "history" of the original woodwork. In other words, how was the woodwork finished over time? Was it originally stained, then painted white 20 years later, followed by a slapdash coat of red? Each layer of old paint or varnish, once removed, will have an effect on your work and determine some of the steps necessary to achieve a realistic and attractive end result when matching new work to old.

For example, the house this mantel was designed for had oak trim with 100 years of finishing history. After several early coats of varnish, someone decided to "brighten up" the place with a coat of kitchen white enamel. Later, someone decided to strip all the woodwork. The drastic but effective method chosen was to apply a caustic stripper in order to thoroughly remove the stubborn white enamel. Sure enough, this treatment obliterated the paint. But the stripped surface was affected in two very visible ways: Because of the oak's porous surface, traces of white paint were embedded in the wood and remained in the nooks and corners of the woodwork. And the caustic chemical nature of the stripper must have reacted with the oak to leave a noticeable yellow cast to the wood. These residual effects were actually attractive. More importantly, they determined the steps I used to match the new mantel to the existing woodwork.

Tip: Before staining or finishing it, apply a solvent like denatured alcohol or lacquer thinner to the face of the cornice to highlight any remaining traces of the saw cuts.

THE FINISH

On this project, I decided to finish the mantel to match original period woodwork in the rest of the house. I determined its paint history by examining and scraping a small sample (see "Re-creating an Old Finish" on p. 77). It may seem like a lot of steps, but you don't have to apply all the glaze coats.

1. Apply a medium brown stain. I used a 50-50 mix of American walnut and black Solar-Lux stains by Behlen. I thinned this mixture by 50 percent with Solar-Lux thinner and applied it to the oak with a brush.

2. Seal the mantel parts with two coats of water-based polyurethane, sanding carefully between coats.

3. Apply a glaze coat. I used McClosky's glazing liquid with raw umber Japan pigment mixed in. This step simulates the normal aging and darkening of the woodwork. After allowing this to dry for about 15 minutes, I wiped

The first step in finishing the mantel is to apply a 50-50 mix of walnut and black Solar-Lux stains.

As a foundation for the glazing and antiquing treatments that followed, I brushed on two coats of water-based polyurethane.

The application of a raw umber glaze imitated the accumulation of dirt, old wax, and smoke stains over time. A glaze also serves to highlight moldings and other unique details of construction.

Next I brushed on a chrome yellow glaze to duplicate the yellow cast I noticed on other woodwork in the house.

the mantel clean, leaving only traces of the dark glaze in the grain and in the corners and recesses of the woodwork.

4. Brush on a thin coat of polyurethane to "tie down" the raw umber glaze and build up a little more gloss.

5. Allow the finish to dry for a day or two, then apply another glaze coat. This time, mix some medium chrome yellow into the glaze and follow the same procedure as with the dark glaze. I left just a trace of the yellow glaze on the oak to brighten up the color and imitate the yellow cast, probably left by the stripper on the original woodwork.

6. Mix some white Japan pigment in some glazing liquid. Add a touch of raw umber to take the "bright edge" off the glaze. Brush on and wipe off as before. This last glaze embedded itself, like the others, into the wood and created the impression I had hoped for.

7. Finally, lightly scrub the mantel with a turpentine-dampened cloth to remove some of the glaze and create highlights and contrast. In other words, don't make an effort to get all the way into the corners and recesses of the mantel. When everything was dry, I applied some amber paste wax to the mantel and buffed it to a pleasing satin gloss.

THE INSTALLATION

Thanks to careful planning, the complete installation took just about 30 minutes. Everything fit together and required only a minimal amount of scribing or adjustment.

1. Strike a level line across the firebox wall, and attach the 1½-in.-square shelf cleats to the front and side walls. The top of the shelf will be ¾ in. higher than the top of the cleat.

2. Position the mantel shelf (with the returns attached) onto the cleats, and screw it to the wall cleats with countersunk trim-head screws into both the bottom and top of the shelf. The top screws will be concealed by the over-mantel; the bottom ones can be hidden behind the brackets.

The final glaze contained flake-white Japan oil pigment. This glaze left the mantel surface looking like it was just stripped of white paint as part of a recent restoration.

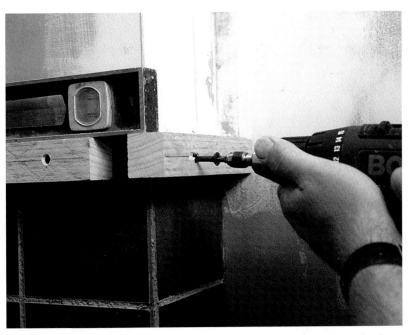

The cleats supporting the mantel shelf are carefully screwed to the fire-place wall with masonry screws.

The back edge of the end panel is scribed with a compass, then planed to fit the wall.

This detail shows the flush frame-and-panel overmantel.

When the shelf and overmantel are fixed, the cornice is dropped into place.

3. Set the overmantel on the shelf and mark the wall along its top edge. Attach two blocks to the wall to support the overmantel—one resting on the shelf and the other below the top line just drawn (see the drawing on p. 68). Check that these two blocks are plumb to each other, and shim as needed.

4. Scribe the back edge of the overmantel return to the wall if necessary, then screw the overmantel to the blocks. (At the top, conceal trim-head screws in predrilled holes through the cornice rabbet. The screw holes along the bottom rail need to be plugged or filled.)

5. Slip the assembled cornice into place, and attach it along the bottom edge with a pin nailer or small brads.

6. Screw the dovetail cleats to the underside of the shelf, and slide the brackets onto the dovetail cleats.

DESIGN VARIATIONS ON THE MISSION STYLE

Because of its longevity the Art and Crafts and Mission styles spawned a wide range of expressions. In other words, there were no definitive Arts and Crafts design parameters. Designs for fireplace mantels could run the gamut from simple rustic designs, hewn from timbers, to tightly structured and carefully crafted versions that resembled fine Tudor paneling. These sketches provide alternative Mission period designs to consider.

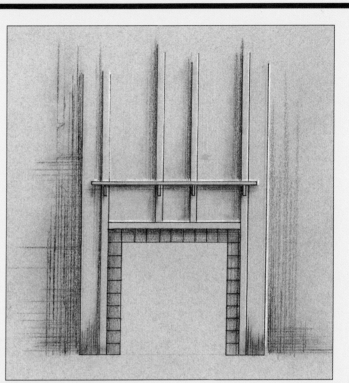

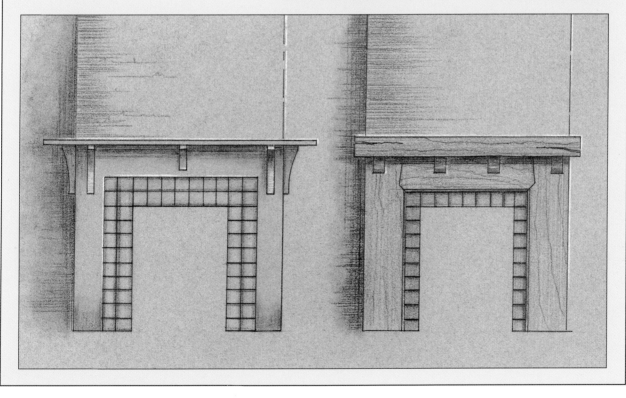

THE MACKINTOSH MANTEL

This mantel is based on the architectural woodwork of The Hill House, designed by the Scottish architect Charles Rennie Mackintosh and built from 1902–1904 at Helensburgh, near Glasgow, Scotland. Although there is a mix of both light and dark furniture and fixtures in the Hill House, the most striking creations are made of oak, stained dark (almost black) with the coarse grain of the wood clearly showing through the finish.

I wanted something that conveyed a specific point in time (the turn of the 20th century), when new directions in design were being explored. Mackintosh's work was symbolic of that period and yet on the cutting edge. His material (oak) and finishes (dark and sedate) were familiar during this period, but his shapes weren't static or predictable. They were dynamic and often mimicked nature. My idea was to design a mantel that was unmistakably inspired by the entire body of Mackintosh's work, but one that could stand on its own merits.

The two most distinctive features of this mantel are the curved frieze board and mantel shelf. They give the mantel a luxurious appearance that plays well against the flat pilasters. But the curves created some practical challenges as well. First, the frieze board, made of ¾-in. oak plywood, is bent over a curved skeletal structure. Second, the mantel shelf tapers gradually to a wafer-thin edge. Such a dramatically overhanging top is a signature feature of Mackintosh's designs.

Mackintosh Mantel

THE DYNAMIC SHAPES AND SUBTLE CURVES in this mantel are characteristic of Mackintosh's designs for the Hill House, near Glasgow, Scotland. The curved frieze, the tapered pilasters, and the compound form of the mantel shelf each provides unique challenges to produce. But the result is a unified composition from a master of architectural design.

PLAN VIEW

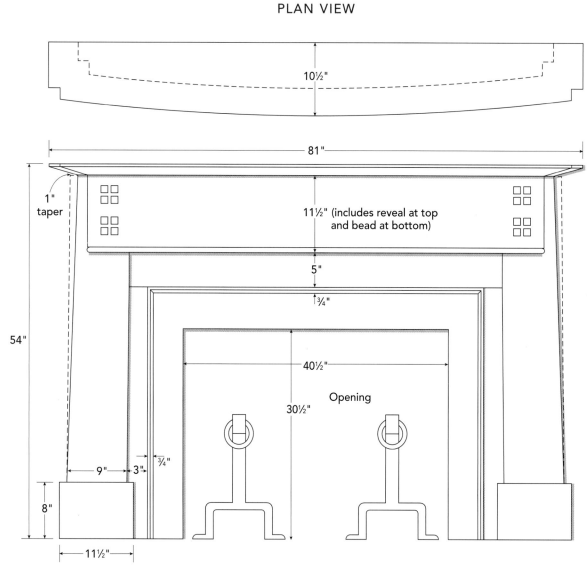

FRONT VIEW

BUILDING THE MANTEL STEP-BY-STEP

This mantel may be the most difficult in the book to build, partly because of the veneer work, and also because of the curved frieze board and mantel shelf. But like most of the other mantels, it's constructed from several smaller assemblies: The foundation is a typical inverted U-shaped assembly; the plinth blocks are made from doubled-up plywood with a return block for scribing; the pilasters are tapered on their outside edges; the frieze board is a curved torsion box sheathed with a bent piece of plywood; and the curved mantel shelf is composed of two plywood decks with beveled solid-wood pieces forming the edge.

MAKING THE FOUNDATION

As the drawing at right illustrates, you can save on the primary oak plywood by joining shop-grade birch plywood to the oak panels. The birch will be covered by the pilasters, and only the oak will be exposed.

1. Cut the plywood parts of the foundation to size. If you're combining a lower-grade plywood with the oak plywood as I did, join the two slightly oversize sections together with biscuits first, then make the final cuts.

2. Lay out and cut the biscuit joints in the foundation boards.

3. Assemble the foundation boards. Make sure the opening is square and the assembly is flat. Screw a temporary batten across the base of the opening for support.

4. Rout a bead-and-quirk molding to trim the interior of the foundation, but don't attach it until the installation.

CHOOSING MATERIALS

This mantel is built almost entirely of ¾-in. quartersawn oak plywood and veneer. Because of the dark finish, it doesn't matter whether you use red or white oak, but straight-grained material is most characteristic of the period. I did use small quantities of ¾-in. shop-grade birch plywood, and some solid pine blocking, in areas that don't show.

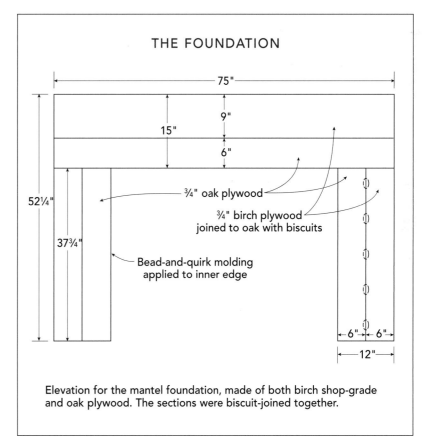

THE FOUNDATION

Elevation for the mantel foundation, made of both birch shop-grade and oak plywood. The sections were biscuit-joined together.

MAKING THE PLINTH AND PILASTERS

Making the plinth blocks

I made these plinth blocks the same way I made those in the simple Federal mantel—by laminating two pieces of stock to get the needed thickness. In this case, though, there's an added return block to add even more depth to the plinths.

1. Cut the plinth block pieces slightly over-size. Use the quartersawn oak for the face pieces, and any ¾-in. plywood for the back pieces.

2. Saw or rout a pair of grooves lengthwise across the inside faces of the panels, and cut splines to fit the grooves.

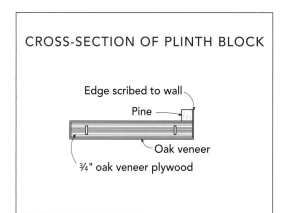

CROSS-SECTION OF PLINTH BLOCK

Edge scribed to wall

Pine

Oak veneer

¾" oak veneer plywood

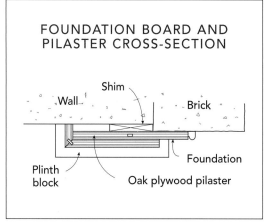

FOUNDATION BOARD AND PILASTER CROSS-SECTION

Shim

Wall

Brick

Plinth block

Oak plywood pilaster

Foundation

3. Glue up the plinth block laminations. When the glue is dry, recut the blocks to final dimensions.

4. Add the ¾-in. by ¾-in. return blocks to the back outside edge of each plinth assembly.

5. Veneer the top and two side edges of each plinth with quartersawn oak veneer. Use the iron-on technique described in the sidebar "Applying Veneer" on p. 92.

Making the pilasters

The pilasters are made of the same ¾-in. quartersawn oak plywood used on the rest of the mantel. The inside edges are plumb and cut square and are veneered after the pilasters are assembled. The outside edges taper from top to bottom by 1 in., giving the mantel a look that is characteristic of the period. The sequence for building the pilasters is shown in the drawing on the facing page, but there are a couple things to keep in mind. Because the pilasters wrap around the foundation and tie into the wall, I made the returns on that side a little long for scribing to the wall at installation. Position the groove for the splines closer to the inside of the edge—if it's centered, it will risk breaking through the face. Also, you need to veneer the inside edge of the pilasters, which I did after assembling the face and return pieces.

MAKING THE CURVED FRIEZE

The curved frieze board is the most prominent feature of the Mackintosh mantel. You'll need to start with a full-sized template showing both the frieze board and the mantel shelf.

Making a template for the mantel top

The mantel top edge curves a scant 1¾ in. from the center to the outside corner. But it's not a true radius—it's flatter in the middle, forming a very subtle curve. (The drawing on p. 88 shows the curve laid out on a 6-in. grid.)

THE TAPERED PILASTERS

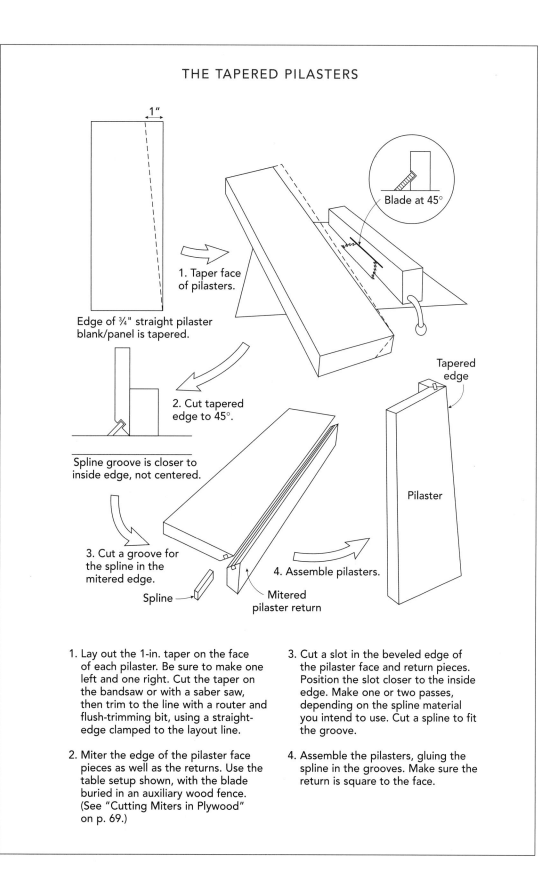

1"

Edge of ¾" straight pilaster blank/panel is tapered.

1. Taper face of pilasters.

Blade at 45°

2. Cut tapered edge to 45°.

Spline groove is closer to inside edge, not centered.

3. Cut a groove for the spline in the mitered edge.

Spline →

4. Assemble pilasters.

Mitered pilaster return

Tapered edge

Pilaster

1. Lay out the 1-in. taper on the face of each pilaster. Be sure to make one left and one right. Cut the taper on the bandsaw or with a saber saw, then trim to the line with a router and flush-trimming bit, using a straight-edge clamped to the layout line.

2. Miter the edge of the pilaster face pieces as well as the returns. Use the table setup shown, with the blade buried in an auxiliary wood fence. (See "Cutting Miters in Plywood" on p. 69.)

3. Cut a slot in the beveled edge of the pilaster face and return pieces. Position the slot closer to the inside edge. Make one or two passes, depending on the spline material you intend to use. Cut a spline to fit the groove.

4. Assemble the pilasters, gluing the spline in the grooves. Make sure the return is square to the face.

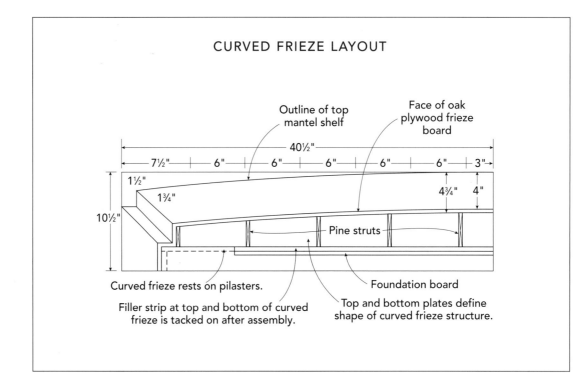

CURVED FRIEZE LAYOUT

Outline of top mantel shelf

Face of oak plywood frieze board

40½"

7½" — 6" — 6" — 6" — 6" — 6" — 3"

1½"

1¾"

4¾" 4"

10½"

Pine struts

Curved frieze rests on pilasters.

Foundation board

Filler strip at top and bottom of curved frieze is tacked on after assembly.

Top and bottom plates define shape of curved frieze structure.

This detail shows the curved frieze in position.

Once you make a template for the mantel top, you can use that as a guide for making two more curved parts—the mantel sub-top, and the curved plates that form the top and bottom of the curved frieze structure.

1. Cut a piece of scrap plywood so it's slightly larger than the finished mantel shelf. Starting at the center, mark lines across the template to form the 6-in. grid pattern. (The first line is 3 in. from center.) Then mark points along the lines to form the curve of the shelf edge.

2. Drive a fourpenny nail or finish screw at each of these points.

3. Hold a thin spline of wood (¼ in. thick by 1 in. wide, for example) against the nails and trace against the spline to form the curve.

4. Make the second template for the bottom half of the top. (Use this to shape the ¾-in.-thick bottom layer of the mantel top.)

Building the frieze board form

The form is a skeletal frame or torsion box made up of a curved bottom and top plate, and several vertical struts.

1. Cut the top and bottom plates, using the template as a guide.

2. Cut the vertical struts all to the same length.

3. Lay the top and bottom plates flat and back-to-back. Mark the spacing of the struts along the back edge of the plates. The spacing is not critical, but approximately every 8 in. on center will work fine. Then strike lines across the plates, square to the back edge, to mark out the position of the struts.

4. Rip the struts to their varying widths according to the layout on the plates.

5. Nail or screw the plates to the struts. Measure across the diagonals at the back of the form to make sure the assembly is square.

Adding the kerf-bent frieze panel

1. Position the frieze board form on the assembled mantel so it's centered on the pilasters. As evident in the center photo, there is a gap between the form and the foundation, which can be filled by tacking on a ¾-in. by ¾-in. strip that will fall between the pilasters.

2. Rip the oak frieze board to width. It should be ⅜ in. less than the height of the form, allowing a step at the bottom for the bead-and-quirk molding.

3. Crosscut one end of the plywood frieze board at a 45-degree angle.

4. Clamp the mitered end onto the form in its final position. With moderate pressure the plywood should bend over the form enough so you can mark the location of the miter cut on the other end. Then make that second miter cut.

5. Cut a series of shallow kerfs into the back of the plywood (across the width) to help it bend easier (¼ in. deep and every 2 in. should work fine). Fill the kerfs with a paste made of sawdust, glue, and water, then bend the kerfed plywood over the curved form and staple it to the frame. Why all the bother filling the kerfs? Sometimes the kerfs in a bent panel will telegraph through the face and be visible after the finish is applied, or after the mantel is installed. Also, the hollow kerfs present weak

Nail or screw through the top and bottom plates into the vertical struts. The skeletal form supports the kerfed oak plywood frieze board.

Lay the curved form on the foundation board to check for fit and position. Add a ¾-in. by ¾-in. strip to the top and bottom plate to fill the gap between the plates and the foundation.

The plywood frieze board is kerfed on the back side, then filled with a mixture of sawdust, glue, and water in order to strengthen the curved panel just before it's nailed to the form.

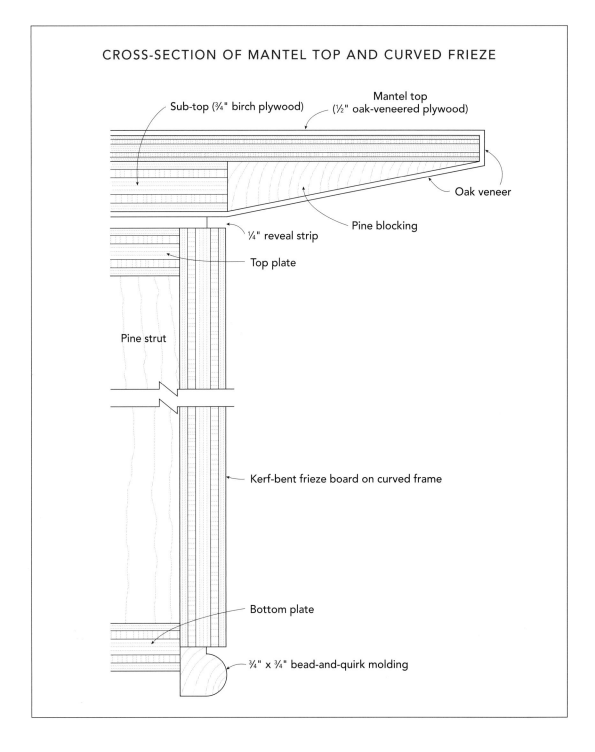

CROSS-SECTION OF MANTEL TOP AND CURVED FRIEZE

Sub-top (¾" birch plywood)

Mantel top (½" oak-veneered plywood)

Oak veneer

Pine blocking

¼" reveal strip

Top plate

Pine strut

Kerf-bent frieze board on curved frame

Bottom plate

¾" x ¾" bead-and-quirk molding

points all along the face of the panel that could crack if struck by a wayward andiron or chunk of firewood. Lastly, the exposure to high heat levels could stress the curved panel and lead to cracking years later. Filling the kerfs is low-cost insurance against these risks.

Finishing off the curved frieze board

1. Cut the short mitered returns for the curved frieze. Note that the grain runs in the short direction. The best way to cut this is to miter a longer piece on the table saw, then

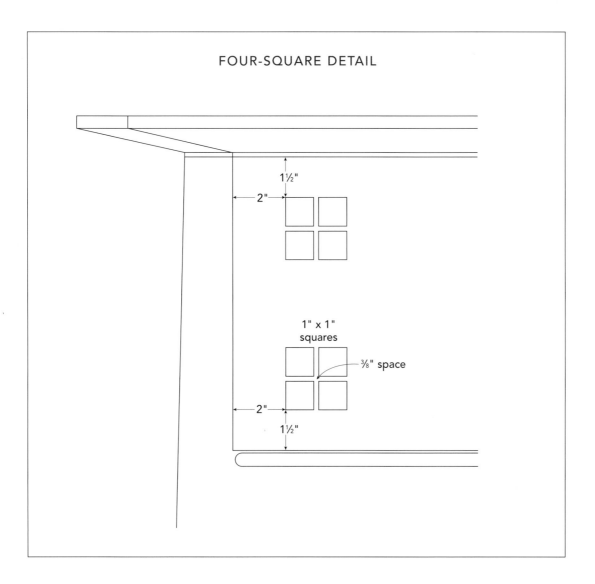

FOUR-SQUARE DETAIL

1½"

2"

1" x 1"
squares

⅜" space

2"

1½"

make the square cut to release the return from the larger piece.

2. Glue the mitered returns in place, and secure them with a few brads if necessary.

3. Nail the bead-and-quirk molding against the bottom edge of the frieze board. The molding is thin enough, and the curve of the panel slight enough, that a straight molding will bend easily around the curve. Miter the ends of the curved molding and add the short returns.

4. Make and attach a spacer onto the top of the frieze assembly to create a neat reveal between the frieze and the mantel shelf. I cut a piece of ¼-in. plywood shaped to match the top of the curved frieze form, but slightly

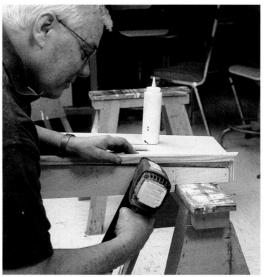

Nail the bead-and-quirk molding onto the bottom edge of the curved frieze board. The molding bends easily around the gentle curve.

In preparation for veneering, yellow glue is applied to the oak veneer and substrate with a small roller. After the glue gets tacky, apply a second coat and then let it dry.

A household iron, set on moderate heat, is used to soften the glue and establish a bond. For small jobs, this method is quick and dependable.

oversize. I tacked this to the top and trimmed it to the front curve with a flush-trimming router bit. Then I routed it back ¼ in. with a rabbeting bit.

5. Make and glue the four-square clusters near the front corners of the frieze. The squares are solid oak, and pre-sanded. I laid out the locations with light pencil lines and simply rubbed them into a dab of glue. A carefully placed pin nail or brad at the center ensures that they'll stay put.

Edging When veneering edges, always cut the veneer slightly larger than the substrate so it will overhang about ¹⁄₁₆ in. on all sides. I coat both the veneer and the substrate with two coats of yellow glue, applied with a brush or small roller. Let the first coat tack up slightly, then apply the second. After allowing the glue to dry thoroughly, I carefully position the veneer onto the substrate. Then, with the iron set on moderate heat, I gently press a hot iron onto the veneer. By slowly moving the iron along the edge, the glue is reactivated. Keep the iron moving constantly to avoid burning the veneer. A moderate amount of pressure bonds the veneer to the substrate. After testing the veneer to ensure a good bond, I trim the excess with either a veneer saw or a small file. When the trimmed edges have been cleaned, I can better inspect and evaluate the joint. Any gaps can usually be closed with a little more ironing.

APPLYING VENEER

I frequently use yellow glue and a household iron to apply veneer to a substrate. It's a low-tech method especially suited to smaller surfaces. Any electric iron will work, but a better one will provide more even heat and do a better job.

Flat surfaces I follow the same steps when gluing larger areas, except that I mist the face side of the veneer with a little water to counteract the veneer's tendency to curl up when it's set down onto the yellow glue.

When the glue on the veneer and substrate are dry, I position the veneer onto the substrate and, starting from the center, iron out toward the edges. Be sure to work the entire surface. Any spots that are missed will show up later as blisters or bubbles. Fortunately, with this technique, if you find any bubbles before applying your finish, they can be ironed down—even days later!

MAKING THE MANTEL SHELF

There were a couple of aspects of the mantel shelf's design that complicated its construction. The first was the wide bevel that runs along the underside of the shelf's edge, and the other was the notch at the ends of the shelf that requires short sections of the beveled surface to be mitered. I decided to make the shelf out of two layers of plywood, add solid-pine wedges to form the beveled edge, and then veneer the entire beveled edge along with the leading edge of the shelf.

Building the shelf

The shelf has two layers, ½ in. for the top, and ¾ in. for the bottom, which is set back from the top 4 in. The 4-in. ledge is filled in with pine blocks milled to a precise bevel. The beveled blocks are added along the front edge up to the first miter at the notch, and then the bevel is veneered. The two small corner bevels, as well as the return bevel, are cut from a longer section that's been veneered already, then assembled one to the next to form the three-part return piece.

1. Cut the ½-in. top and ¾-in. sub-top to size. Using the template, lay out and cut the curved front edges on each piece.

This detail shows the finished foursquare detail.

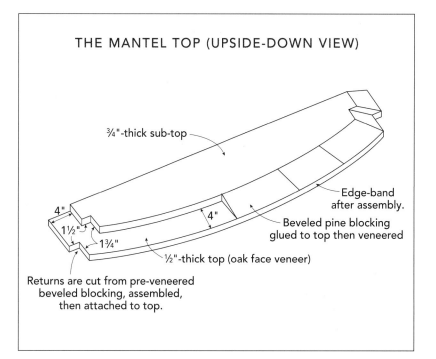

THE MANTEL TOP (UPSIDE-DOWN VIEW)

¾"-thick sub-top

Edge-band after assembly.

4"

4"

Beveled pine blocking glued to top then veneered

1½"

1¾"

½"-thick top (oak face veneer)

Returns are cut from pre-veneered beveled blocking, assembled, then attached to top.

The precise position and angle of each miter is laid out on the plywood mantel shelf parts, defining the small mitered pieces and the returns.

Pine struts that are 4 in. wide were run through the thickness planer on this jig to achieve a uniform taper. The L-shaped jig and its base remain stationary; only the stock moves through the planer.

2. Lay out the exact location of each miter and return on the main top shelf. Then screw the bottom section of the shelf to the top, making sure that the corners of the notches line up.

Making the beveled edge stock

1. Mill the beveled edge stock. (Six pieces ¾ in. by 4½ in. by 20 in. will give you an extra piece or two.)

2. Make a tapering jig for the planer like the one shown in the bottom left photo. It's a simple platform with a ½-in. raised lip on one edge. Position the stock with one edge up on the lip, and send it through the planer.

3. Crosscut the ends of these blocks at 89 degrees to accommodate the curve and ensure a tight joint between each piece.

4. Shape the back edge of each beveled piece to conform to the curved edge of the sub-top. All this takes is several passes with a block plane or bench plane to remove about ⅛ in. of material from the center of each piece.

5. Glue down the tapered blocks along the front edge of the shelf, working from the center toward the ends. The end blocks were cut to 45 degrees according to the layout marks on the plywood.

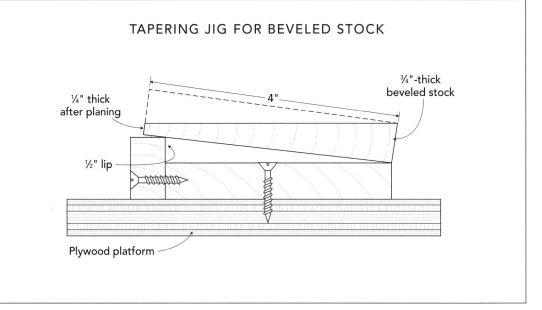

TAPERING JIG FOR BEVELED STOCK

¼" thick after planing

¾"-thick beveled stock

4"

½" lip

Plywood platform

After shaping the beveled edge material to conform to the curved shape of the shelf, clamp and glue the pieces along the entire edge of the shelf up to the notches. Then veneer the beveled edge, allowing the veneer to extend about ½ in. onto the flat.

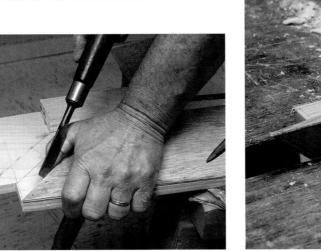

Trim the veneer flush at the ends using a sharp chisel.

6. When all the blocks along the front edge are glued down and trimmed, veneer the beveled surface. Let the veneer extend ½ in. over the seam onto the flat. It will bend easily under the heat and pressure of the iron. Trim the veneer flush with the first miter cut with a sharp chisel.

Adding the mitered returns

Unlike the main sections of beveled edging, the return sections are veneered before cutting and assembling them.

1. Cut the four short miter returns and the two long miter returns from veneered lengths of beveled stock. Cutting the short miters (either a table saw or miter saw will do the job) is another case where the "offcut" piece is actually your workpiece. I cut the first edge of each short miter piece, then position the cut end on a pencil mark instead of a stop, which would dangerously trap the small offcut; then I make the cut. You may need to fine-tune the fit with a block plane.

2. Assemble the three pieces to form the complete return. I glued the miters using simple

The individual pieces forming the corner miters and returns are cut and assembled before being applied to the shelf. Here one of the short returns is being trimmed for a clean miter joint.

The assembled corner miters and returns were fit and glued onto the plywood shelf, and then the front edge was veneered to complete the shelf.

This detail shows the underside of the corner miters.

rubbed joints, keeping the assembly flat on my workbench.

3. Glue down the return assembly to the top.

Finishing up the shelf

1. Sand or plane the front edge of the shelf so the beveled edge is flush and square with the edge of the top plywood. Use a sharp chisel to clean up the notches.

2. Veneer the front edge. Start with the long center section and work around the notches to the returns.

3. Sand the shelf.

THE MACKINTOSH FINISH

The finish used in the Hill House fixtures and furniture would be perfect for this mantel. It was a dark chocolate brown stain applied over oak, probably finished with shellac or varnish, then waxed. However, in photos of the original mantel, this finish didn't appear to wear well. The color was worn off the sharp edges, and nicks to the surface revealed the raw oak below. But the casualness of the surfaces was appealing. The finishes Mackintosh used were straightforward and seemed almost improvised. They certainly weren't anything like the slick, smooth surfaces we've come to expect. I had to come up with a finish that possessed the charm of the original but wore a little better and aged more gracefully.

Preparing the surface

In the Hill House mantel, the oak surfaces were not filled and smoothed before finishing. The oak was slightly coarse; the oak grain, raised and visible. This crude surface enhanced the Arts and Crafts origins of Mackintosh's work and gave the wood an unrefined appearance in keeping with other natural materials used in the house, such as slate, stucco, and fieldstone.

1. Coat the wood surfaces with a light mist of water and let the wood dry.

2. Scrub the wood gently with wire brushes in the direction of the grain. The resulting sur-

Before staining, go over the surface gently with a wire brush to raise the grain. This treatment produces an interesting surface that highlights the coarseness of oak.

The assembled frieze is shown after staining. I mixed three colors to get the exact quality I wanted.

face should be only slightly rough, but nicely raised.

3. Stain the wood. (I used a mixture of 40 percent American walnut, 40 percent cordovan mahogany, and 20 percent jet black Solar-Lux stains from Behlen.)

4. After 24 hours, gently buff the surfaces with 00 steel wool to remove any excess stain, to knock off any superficial roughness, and to slightly "shade" the parts.

5. Apply a sealer coat of shellac to the mantel. This will give you a preview of the final color but is still malleable. I decided that the color was too dull, the brown was too thick. So I souped it up with a wash made of bulletin red Japan pigment mixed with turpentine. This gave me just the right warm shade of brown, and after the wash dried, I applied two more

ADDING A GLAZE COAT

The appearance of the Hill House woodwork so favorably impressed me that I wanted to re-create its well-worn surface patina and antique color. Over time woodwork accumulates dirt and old wax in the corners and crevices, highlighting the details. The passage of time also replaces the original uniform color with one that is variegated, muted, and slightly faded, but much more interesting. To imitate this look, I decided to glaze the mantel. This involves the use of a prepared clear glaze coat (I use McClosky's) with a small amount of black or raw-umber Japan pigment added to it. The glaze is a slow-drying vehicle that permits the application of an additional color to highlight the project or slightly alter its color. I usually glaze the project after installation; but because the open-pored surface of the mantel might absorb and retain more glaze than intended, I decided to glaze the project in the shop, where I would have more control over the process.

coats of orange shellac, rubbing between each coat. After rubbing out the last coat of shellac, I applied a clear paste wax and buffed the mantel with a soft rag.

This detail shows the pilaster set on the plinth as viewed from the firebox side.

INSTALLING THE MANTEL

1. Position the foundation against the wall, and add shims if needed to get it level and plumb. Screw the foundation to the wall.

2. Scribe the plinth blocks to the wall and floor as needed, and secure them to the foundation with a spot of glue and a few finish nails.

3. Set the pilasters onto the plinth blocks and scribe them to the wall. Measure the distance between the pilasters to ensure that the curved frieze board assembly will fit correctly. Screw the pilasters to the foundation with 1¼-in. nails, or narrow-crown staples if you have a staple gun.

4. Screw support blocks for the curved frieze to the pilasters with #8 by 2-in. wood screws.

5. Lift the curved frieze onto the support blocks, and screw into the support blocks from above and below.

6. Scribe the back edge of the mantel shelf to the wall, and attach the shelf to the top of the curved frieze with countersunk drywall trim-head screws. Fill the holes with wax filler.

7. Miter the bead molding, and tack it onto the inside edge of the foundation opening.

The mounting blocks for the curved frieze are attached to the pilasters with #8 by 1½-in. wood screws.

With the support blocks in place, the curved frieze is set into position. When everything checks out, the frieze board is screwed, top and bottom, to the mounting blocks.

MACKINTOSH SKETCHBOOK

Although Mackintosh disassociated himself from the Art Noveau movement, which was characterized by sinewy flowing curves and realistic natural forms, his own work was influenced by European designers like Gaudi, Hort, and Henry Van der Velde.

Mackintosh employed great fluid curves and abstracted natural forms; but unlike his predecessors, he combined or contained these with harder, more structured elements.

His work is powerful, disciplined, and tightly organized. His furniture and buildings clearly contributed to the the Art Deco movement of the 1920s.

These sketches explore the combination of curves with structured grids, squares, and straight lines. I was looking for some kind of design signature that would be immediately recognized. I found it in the four-square motif that was used on the fireplace mantel.

Victorian Mantel

Within the Victorian period (roughly 1830 to 1900), there were several distinct styles (Eastlake, Art Noveau) and numerous trends (gothic, rococo revival) that influenced interior furnishings. Like the architecture of the time, Victorian interiors often displayed an assortment of ornate designs and patterns juxtaposed alongside one another. We think of Victorian furnishings generally as gaudy, richly carved, and overly embellished.

The industrial revolution of the mid-19th century had a notable influence on the Victorian style. Instead of being hand-carved by individual woodworkers, much of the ornament on Victorian furniture could be mass-produced with the aid of machinery. The goal was to keep the hand-detailed look but gain the economy of mass-production, and this mantel succeeds in that mission.

Following various antique examples, I designed this mantel to be built upon a typical inverted U-shaped foundation. Also typical, each ornamental element was constructed independently; the pilasters, brackets, and moldings were then mounted to the foundation so the completed mantel could be installed as a single finished unit. Incidentally, this method of construction has made the removal and reuse of old mantels easy. So easy, in fact, that for years there has been a flourishing market in stolen mantelpieces—taken right out of occupied homes, while the owners were away for the weekend.

Victorian Mantel

A VISUAL SMORGASBORD OF SHAPES AND PATTERNS gives this Victorian mantel its jaunty appeal. Each element is formed separately, principally from solid wood, then applied to a simple flat foundation.

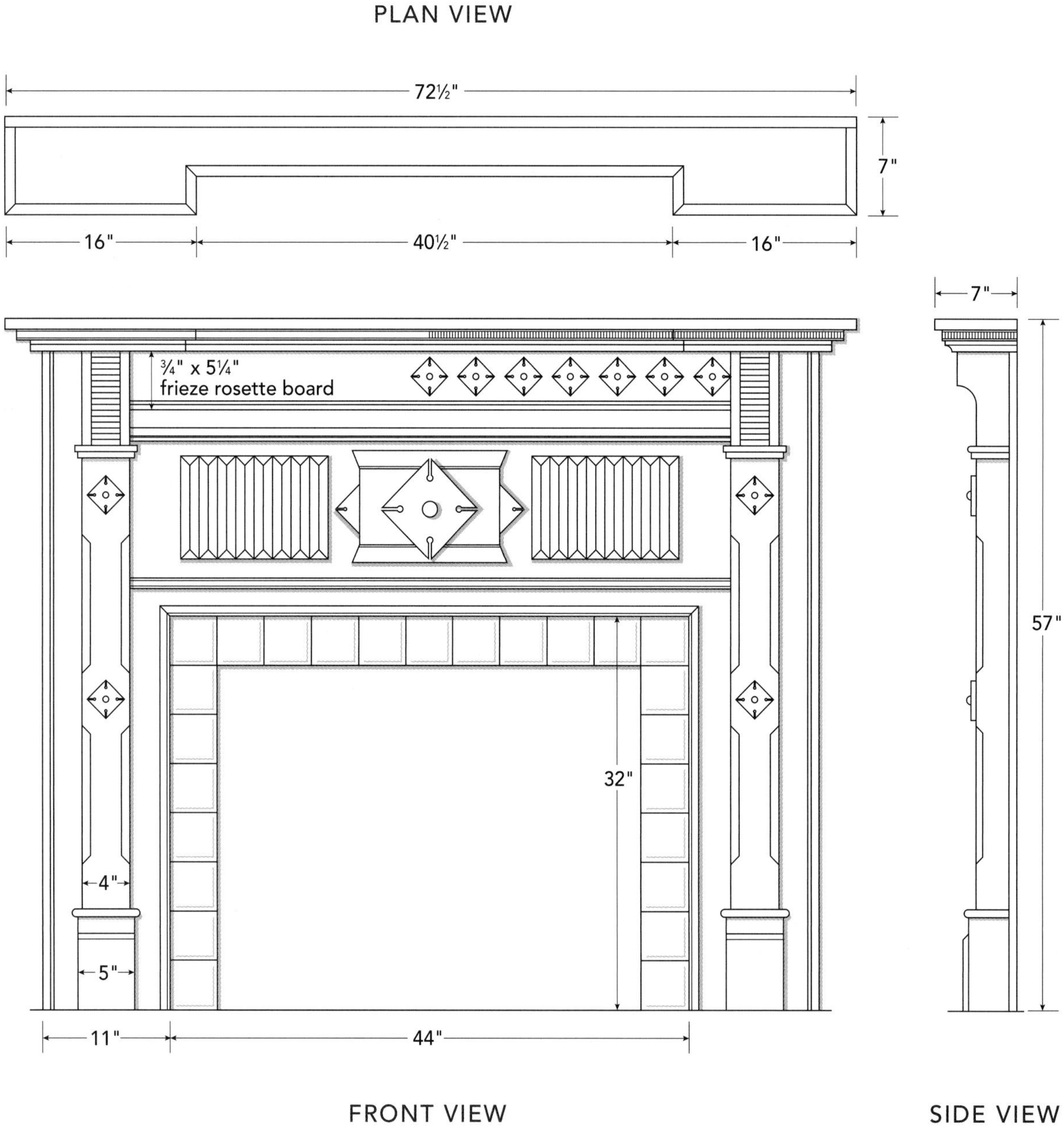

PLAN VIEW

72½"

7"

16"   40½"   16"

¾" x 5¼"
frieze rosette board

32"

4"

5"

11"   44"

7"

57"

FRONT VIEW

SIDE VIEW

BUILDING THE MANTEL STEP-BY-STEP

Much of the decoration on this mantel consists of straight molding stock, laid in a conventional linear fashion or cut up and reconfigured into particular designs and patterns on a foundation board. There are narrow reeds employed on the brackets and as edging on the middle mantel shelf. Triangular moldings are cut into strips with beveled ends and assembled to form the frieze panels. The rosettes are made on the drill press and table saw with a simple holding jig. I started at the center with the frieze board components and worked my way out to the rest of the parts.

BUILDING THE FOUNDATION

I built the foundation for this mantel from ¾-in. mahogany plywood. But instead of resting the lintel (horizontal) section on top of the column (vertical) sections, I extended the columns to the underside of the mantel shelf and set the lintel between the columns for a neater appearance. Either way has structural integrity, and I've come across many examples of each type.

1. Cut the foundation parts to size.

2. Join the sections together with biscuit joints.

After joining the parts of the foundation with biscuits and clamping the assembly, check the glue-up for flatness with a straightedge.

3. Clamp the horizontal section between the two vertical sections. Check the assembly for square, and lay a straightedge across the joints to ensure that the assembled surface is flat.

4. The foundation is trimmed on both the outside and inside with a bead-and-quirk molding. You don't need it until installation, but you can cut and shape it now.

CHOOSING MATERIALS

Though not a native wood to either North America or Europe, mahogany was used extensively in the Victorian period. It's more easily carved than some native hardwoods and can be stained a wide range of colors. Unlike oak (another Victorian favorite), mahogany has a subtle grain pattern that doesn't compete with the design elements in this mantel.

For the wide foundation panels, I used ¾-in. mahogany plywood so I could avoid having to glue up wide panels. While this sheet material is more and more readily available in general-purpose home centers and the like, you'll have to purchase solid mahogany for the moldings and the other elements of the mantel at a hardwood lumber supplier.

MAKING THE FRIEZE COMPONENTS

On this mantel the frieze board is composed of two panels flanking a central cartouche. The panels are made up of pyramid-shaped strips laid side-by-side. This unusual configuration produces a strong vertical pattern that throws some very dramatic shadows across the mantel frieze. It is an interesting effect that I first encountered on a pair of Victorian entry doors in upstate New York. A large rosette is centered on the cartouche, and a row of smaller rosettes runs under the mantel shelf.

Making the pyramid molding

The pyramid molding was produced on the table saw with a simple jig that supported 1⅛-in.-square stock at a 45-degree angle while

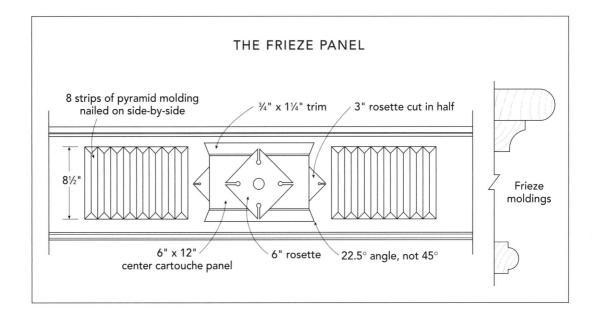

THE FRIEZE PANEL

8 strips of pyramid molding nailed on side-by-side

¾" x 1¼" trim

3" rosette cut in half

8½"

Frieze moldings

6" x 12" center cartouche panel

6" rosette

22.5° angle, not 45°

Mahogany strips, milled to 1⅛ in. square, were cut into long pyramidal molding strips on this shop-made jig. The completed molding strips assembled eight-in-a-row form the side frieze panels.

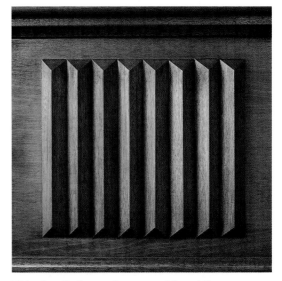

This detail shows the pyramid moldings applied to form a panel.

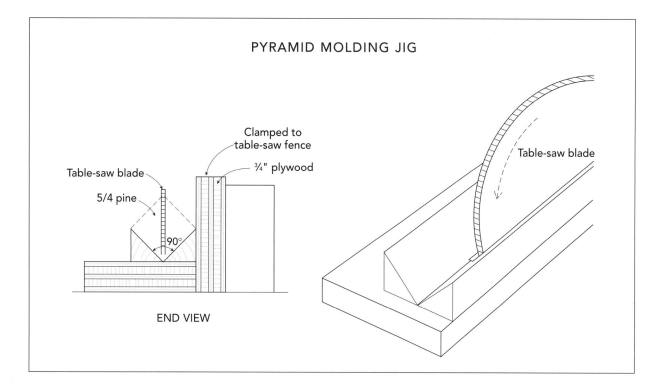

PYRAMID MOLDING JIG

Table-saw blade

5/4 pine

Clamped to
table-saw fence

¾" plywood

90°

END VIEW

Table-saw blade

keeping the saw blade at 90 degrees. The roughsawn sides are placed against the mantel foundation.

1. Rip the square stock into two triangular pieces on the table saw, using a jig like the one shown in the drawing above. Use a push stick for safe, controlled cuts.

2. Cut the molding pieces to length.

3. Bevel the ends of each piece on the miter saw with the blade set at 45 degrees. A good sharp blade will leave a smooth cut, but I shaved the cut ends with my miter trimmer for a super-smooth surface.

Making the square rosettes

I wanted a small distinctive design element that was easy to produce in small quantities on shop machinery, but that suggested the hand-carved rosettes used during the Victorian period. In my mantel design, I placed a band of these rosettes across the frieze between the brackets and a couple on each pilaster. These simple yet attractive squares set on the diagonal were easy to make on the table saw and drill press. (Be sure to make a couple extra

A miter trimmer cleans up saw cuts on mitered moldings—especially valuable here, where the cut is exposed. This 100-year-old device, once deemed obsolete by the miter saw, is making a big comeback.

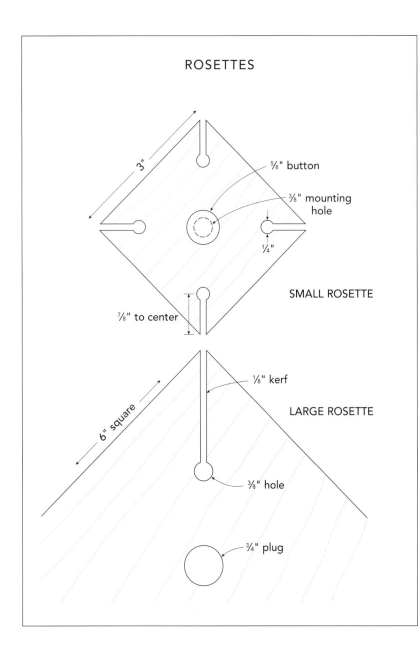

ROSETTES

3"

⅝" button

⅜" mounting hole

¼"

⅞" to center

SMALL ROSETTE

6" square

⅛" kerf

LARGE ROSETTE

⅜" hole

¾" plug

A shop-made jig precisely positions each rosette for drilling holes—here the center countersink for mounting the rosette.

The same jig is used on the table saw for the kerf cut. The stop block clamped to the fence prevents the rosette from being cut in half.

rosette blanks—one of them is cut in half and used on the center cartouche.)

The key to producing these rosettes was to build a jig that would support each 3-in. square during two different operations

1. Clamp the rosette holder on the drill press, then drill a ¼-in. hole ⅞ in. from each corner.

2. Reposition the rosette holder and drill a ⅜-in.-deep by ⅜-in.-diameter countersink at the center of each blank for a mounting screw and button plug. Reserve one of the rosettes

for the center cartouche—this one doesn't get the center countersink.

3. Cut a saw kerf from each corner to the ¼-in. hole.

4. Repeat the process for making the large rosette for the center cartouche.

Making the center cartouche

Here I wanted something bold and architectural, nothing too fussy or detailed. So I chose to repeat the rosette design, only larger, and mount it on a plywood panel.

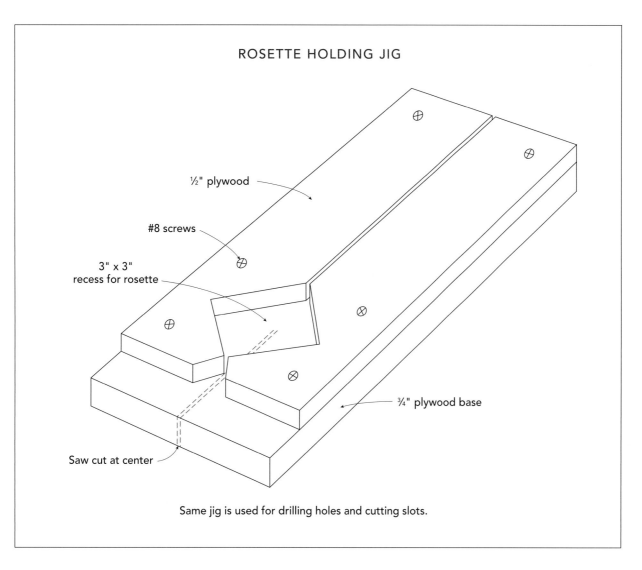

ROSETTE HOLDING JIG

½" plywood

#8 screws

3" x 3"
recess for rosette

¾" plywood base

Saw cut at center

Same jig is used for drilling holes and cutting slots.

1. Cut the plywood panel to size.

2. Rout a small rabbet along the long edges of the panel. This creates a slight reveal when the cartouche is assembled.

3. Rip the small rosette in half, using the holding jig.

4. Cut the top and bottom trim pieces to size with their ends beveled at 22.5 degrees.

5. Cut biscuit slots to join the top and bottom trim and half-rosettes to the panel.

6. Assemble the center cartouche. The center hole will be used to mount the rosette to the foundation, and then capped with a round mahogany plug.

This detail shows the rosette applied to the pilaster.

ROSETTE VARIATIONS

While I was designing the machine-made rosette, my sketches generated several other ideas for rosettes that could have worked as well. Each one met my criteria for a machine-made ornament that would be Victorian in spirit and give the mantel a distinct 19th-century appearance.

Like the rosette in the production sequence pictured, any of the other illustrated designs could be produced in a similar manner, using a simple shop-made jig. Also shown is a traditional Victorian design for a carved rosette.

CARVED VICTORIAN ROSETTE

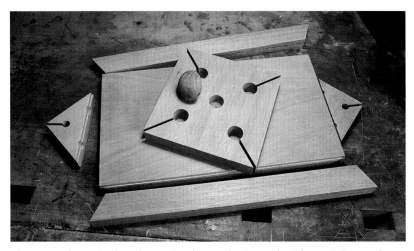

The central cartouche plaque is plywood trimmed with straight stock at the top and bottom, and half-rosettes on the sides, joined together with biscuits. The large rosette is screwed to the plaque, then plugged.

7. Rout the two horizontal moldings that frame the frieze.

MAKING THE MANTEL SHELVES

For this vigorous Victorian mantel, with all of its applied ornament, brackets, and heavy pilasters, a graduated three-tier shelf was necessary to maintain some visual balance. Each tier was made of ¾-in. plywood, then edged with a solid-wood component. The top shelf is edged with ¾-in.-square solid mahogany and left that way. The middle shelf is decorated with the reeded molding. And the bottom shelf gets a cove molding glued to the edge.

The corner of each mantel shelf tier must line up to ensure proper alignment of miters and uniform position of each tier. Here lines scribed at 45 degrees make registration of each tier easier. (Note that the middle shelf is birch, not mahogany, because only its mahogany edge detail will show.)

Cutting the shelves

I could have cut the square recess in the front of each shelf, but it would have been tough to get good clean inside corners. Instead, I ripped the shelf into two pieces, crosscut the front piece to yield the two steps on either end, then glued them back together, keeping the grain intact.

1. Cut the shelf blanks about ¼ in. longer and wider than the final dimensions.

2. Rip each shelf to form the center notch, but mark the ends that will be reglued in order to maintain the grain match.

3. Crosscut the two front end pieces to size for each shelf.

4. Cut biscuit joints, and reglue the front end pieces to the back half of each shelf.

5. Trim the shelves to their final dimensions. After cutting the shelves, lay them upside down and scribe a line at 45 degrees from the corners. Stack the shelves to make sure the mitered marks line up.

Making the beaded molding

A lot of fussy work goes into this small but important molding detail. Oriented vertically

MANTEL SHELF

¾" x ¾"

⅛" x ⅝" solid mahogany

¼" x ½" bead strips on ¼" ply backer

¾" x ¾" cove

¾" plywood

Segments of the reeded stock that are ½ in. long are laid side-by-side on a plywood strip. The matrix strip supports and reinforces this complicated built-up molding strip.

here, the reeds on the middle shelf edge act as a visual counterpoint to the same detail on the bracket faces.

1. Rout the three-reed profile onto the edge of the molding stock.

2. Rip the reeded edges from the stock—it should be ¼ in. thick.

3. Cut the reed stock into ½-in.-long segments on the bandsaw, and glue these short pieces onto 24-in.-long plywood strips to form the beaded matrix strips.

4. True up the glue joint with a sanding block.

5. Glue the matrix strips between the thin solid-wood strips (⅛ in. by ⅝ in.). Sandwich the pieces together on a flat surface, and clamp the assembly with handscrews.

Trimming the mantel shelves

1. Plane the square stock for the top shelf, and shape the cove molding for the bottom shelf.

2. Cut and glue the edge treatment onto each shelf. I "walked" my edging around the shelves, starting at the left corner and working my way around to the right side. I took my time cutting and fitting each segment to ensure a tight and attractive fit. Remember, the shelf is at eye level and will be subject to some close scrutiny, so don't settle for anything less than a perfect job.

3. I used clamps wherever they would supply necessary pressure without damaging the molding. To protect the delicate edges of the middle shelf, I used a continuous block between the clamps and the workpiece. On the bottom shelf, I secured the small cove molding with masking tape.

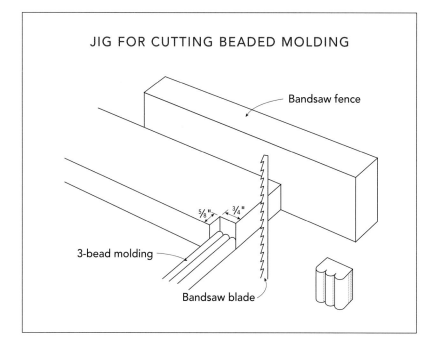

JIG FOR CUTTING BEADED MOLDING

Bandsaw fence

3-bead molding

⁵⁄₈" ¾"

Bandsaw blade

Profiled moldings might be difficult to attach with conventional clamps. Here a delicate cove molding is held to the plywood edge with masking tape.

Thin strips of mahogany are clamped to the reeded strip with handscrews.

The top mantel shelf is right at eye level, so seams and miters must be clean and precise.

MAKING THE BRACKETS

Victorian architecture concentrated a large part of a house's decoration at the eaves of the house. The viewer's eye very naturally travels to the juncture between the façade and the roof and is rewarded with a lively band of gingerbread, brackets, and Victorian filigree. I wanted to imitate the playfulness found in vernacular Victorian architecture and thought this would be a good place to do so. In this design, the brackets support the shelf and crown the pilasters. They are the most dynamic element

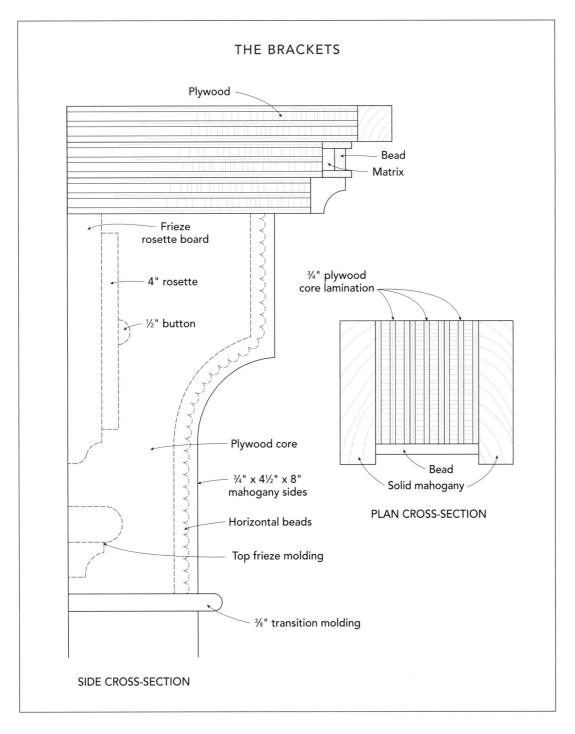

THE BRACKETS

Plywood

Bead

Matrix

Frieze rosette board

4" rosette

½" button

¾" plywood core lamination

Plywood core

¾" x 4½" x 8" mahogany sides

Horizontal beads

Top frieze molding

⅜" transition molding

SIDE CROSS-SECTION

Bead

Solid mahogany

PLAN CROSS-SECTION

on the mantelpiece and unite the pilasters, running vertically and anchored to the floor, with the shelf, which spans the hearth and crowns the mantel.

The brackets are made by sandwiching a 2½ in.-thick plywood center section between two pieces of ¾-in. solid mahogany. The out-line of the side sections projects ⅜ in. past the center section, creating a shallow setback that is decorated with short reeds laid horizontally.

1. Make a full-sized pattern for both the core pieces and the sides of the bracket.

2. Cut the sides of each bracket out of ¾-in. mahogany on the bandsaw. Trim the sides to the pattern with a router and flush-trimming bit.

3. Cut and shape the core pieces in the same way, then glue three pieces together to form the bracket core. Clean up the curved shape on each bracket and center section with a spokeshave and a small sanding block.

4. Crosscut the reed molding to fit on the center section. Glue the reeded strips onto the curve of the interior bracket section with yellow glue. The curve was shallow enough and the reeded sections narrow enough not to require any special clamping methods to hold them to the curved surface of the interior bracket. I just rubbed each reeded section into the glue until the joint developed some suction, then left it to dry.

5. When the reeded center sections were dry, I planed the ends of the reeds flush with the sides of the center blocks. Then I glued the exterior side sections to the interior reeded section and clamped the assembly with handscrews.

Sections of reeded molding that are 2½ in. long are glued to the interior bracket. Once the reeded section is dry and trimmed flush using a smooth plane, the exterior mahogany brackets are glued on.

THE PILASTERS AND PLINTHS

I milled the pilasters and plinths from solid 2½-in.-thick mahogany. If you can't find material this thick, laminate two or more pieces together.

The pilasters

1. Mill the pilasters square and cut them to size.

2. Lay out the stop lines for the two stopped chamfers, and rout them. Make one or two passes to remove most of the material, then make a final light pass to avoid tearout (see the photo at bottom left).

3. Clean up the ends of the chamfers with a rounded sanding block.

Stop the chamfers on the pilasters using clear layout lines for visual guidance.

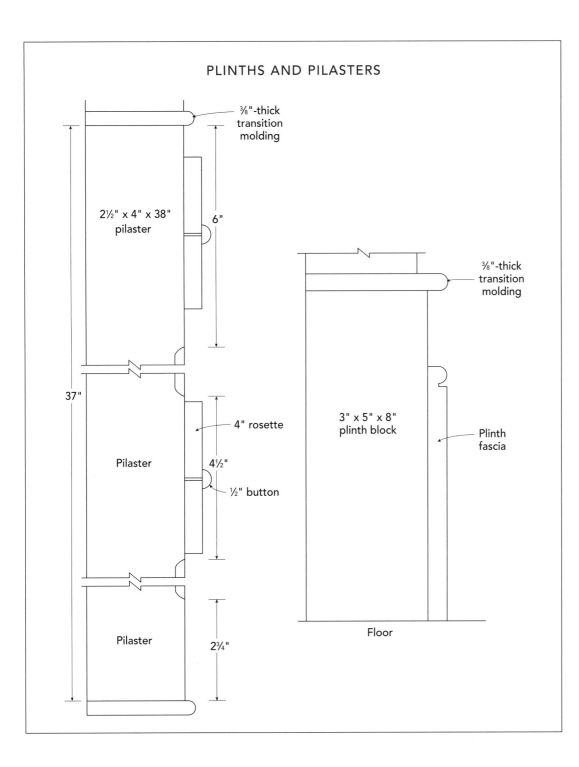

PLINTHS AND PILASTERS

³⁄₈"-thick transition molding

2½" x 4" x 38" pilaster

6"

³⁄₈"-thick transition molding

37"

4" rosette

Pilaster

4½"

½" button

3" x 5" x 8" plinth block

Plinth fascia

Pilaster

2¾"

Floor

The plinths

The plinth is a two-piece assembly—a 2½-in.-thick main block with a ½-in.-thick face piece glued on to create a slight step.

1. Cut the parts to size.

2. Shape the top edge of the face piece, and add the horizontal reveal. (I cut the reveal on the table saw with a sharp crosscut blade. A ⅛-in. router bit will also do the trick.)

3. Glue the face piece to the plinth block. Avoid (or quickly clean up) any squeeze-out along the top edge of the face piece. When the glue dries, block-plane and sand the sides of the plinths.

4. Mill about 4 ft. of the transition molding for both the top of the plinth and the top of the pilasters.

ASSEMBLING THE MANTELPIECE

Most of the mantelpiece can be assembled prior to installation. The only parts that can't be assembled in the shop are the shelves and the bead-and-quirk moldings—they should be scribed to the wall surfaces to conceal any gaps. Wherever possible, I screwed the parts to the foundation from behind; otherwise I used a brad nailer.

1. Center the pilasters on the vertical foundation sections, working from the bottom up. Starting with the heavy plinth blocks, then the pilasters, followed by the brackets, attach each of these components to the foundation from behind with countersunk #8 by 1¼-in. screws.

2. Cut the ¾-in. by 5¼-in.-wide frieze rosette board to fit between the brackets. (This strip receives the rosettes.) Rout a roundover profile into its bottom edge. Screw it from behind to the foundation.

This detail shows the plinth and base of the pilaster.

3. Cut the two frieze moldings to fit, and nail them in place.

4. Draw layout lines to position the pyramid moldings and the center cartouche. Nail the pyramid strips in place one against the next.

5. Screw the center cartouche in place with the screws concealed behind the rosette.

6. Attach the small rosettes to the mantel with #8 screws set into the countersunk hole in the center, then glue a decorative plug into the hole. (You may want to remove a couple of the rosettes when you install the mantel in order to conceal the mantel-mounting screws behind two or more of the rosettes.)

Finishing the mantel

After laying the assembled mantel flat on sawhorses, I stained the mantel with one coat of walnut stain, followed by a light coat of a red wash (Japan pigment suspended in turpentine). When this was dry, I sealed the mantel with a washcoat of orange shellac. The next day I rubbed out the mantel, careful not to cut through the finish on the corners. Then I carefully brushed on two more thin coats of shellac and left it to dry.

INSTALLING THE MANTEL

I installed this mantel with just four masonry screws, placed into ½-in. countersunk holes, and driven directly into the masonry wall behind the mantel. The holes were located in spots that wouldn't easily be seen and could be further obscured with a plug, then a little glaze or wax. (Depending on your conditions, you could conceal screws behind the rosettes on the pilasters and under the mantel shelf.)

Once the mantel was secured to the wall, I fit the two lower shelves onto the mantel and screwed them down into the tops of the brackets with #8 wood screws. I scribed, fit, and finish-nailed the top shelf to the mantel. Lastly, I added the bead-and-quirk molding around the interior of the foundation board and at the side edges of the mantel.

This detail shows the shelf bracket with recessed reeded molding.

Raw-umber glaze is applied heavily with a brush, then rubbed off with a rag.

THE FINISHING TOUCH

I wanted this mantel to look like a pristine architectural gem pulled from some crumbling Victorian mansion—saved from demolition, dusted off, touched up, and given a new life. The subtle discoloration that occurs over time, the accumulation of dust, dirt, and furniture wax in the nooks and corners, and the soft burnished sheen of polished woodwork are all desirable qualities in salvaged or antique wood-work. And the best way to imitate the look of heavy use and loving care is to use a glaze.

Glaze coat is a viscous clear vehicle with added pigment. I mix raw-umber Japan color into the glaze coat for a soft, convincing effect. I made sure to thoroughly mix the ingredients, then I brushed the glaze onto the mantel, almost coating the entire surface and carefully working it into the corners and recesses. After allowing the glaze to set, but not dry, I rubbed most of it off with a turpentine-dampened rag.

The finished mantel overall took on a slightly darker, more subdued color than I expected. In addition to making the mantel look older, the glaze coat also highlights important details, such as the reeded moldings and the rosettes.

A successful glaze coat gives the new mantel the appearance of a salvaged architectural gem.

GREENE AND GREENE MANTEL

The Greene brothers were a pair of architects working in southern California during the early 20th century. They designed houses and everything that went into them, including furniture, lighting fixtures, and of course, fireplace mantels. Influenced by the Arts and Crafts movement that originated in England, their work had an austere quality that employed exposed beams and joinery, protruding pegs, wrought-iron strapping, and braces. These features, which also drew inspiration from Japanese architecture, defined the dimensions and character of a room. Combined with the use of dark muted colors, they created a snug and secure atmosphere.

Though inspired by a mantel in the Blacker house in Pasadena, California, this mantel is based more on the entire body of Greene and Greene interior work. It has a sprinkling of their typical design elements: the use of exotic wood, quarter-rounded corners, rosewood pegs, and the Greene brothers' signature "cloud lift" motif.

Economical in scope and fairly easy to build, the mantel consists of two vertical stiles connected by a horizontal rail. This "H" construction carries two vertical brackets that support the 3-in.-thick mantel shelf. Above the shelf, between the outer stiles, is a flat paneled overmantel. The rails of the overmantel are decorated with the cloud lift motif. And capping the overmantel is a simple flat crown.

Greene and Greene Mantel

THIS MANTEL IS ROOTED IN THE TRADITIONAL FORM of earlier mantels—a wood surround supporting a thick shelf. But it breaks free from that tradition in its stylistic composition. Simple rounded edges, the use of contrasting woods for the square pegs, and the Greene brothers' signature "cloud lift" motif are carefully integrated to form a thoroughly distinctive mantel.

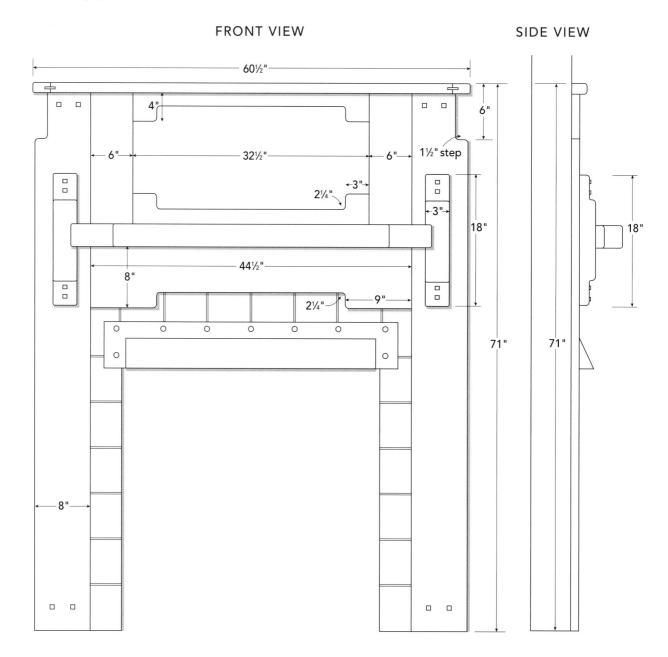

FRONT VIEW

SIDE VIEW

BUILDING THE MANTEL STEP-BY-STEP

Start with the "H" frame, then make the overmantel to fit within the H frame. Next make the brackets, then the shelf. After finishing the mantel and installing it, add the decorative rosewood pegs to conceal the mounting screws.

MAKING THE H-FRAME

1. Plane the solid stock to ¾-in. thickness.
2. Rip the two stiles and the rail to width, then crosscut them to length.

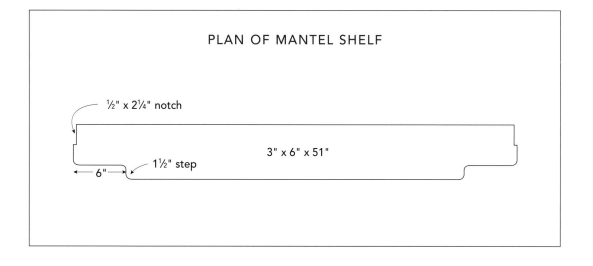

PLAN OF MANTEL SHELF

½" x 2¼" notch

3" x 6" x 51"

1½" step

6"

CHOOSING MATERIALS

Honduras mahogany was the perfect material for this mantel. As solid timber it's available in thicknesses up to 4 in. and as veneer-faced 4 by 8 panels in ¼-in., ½-in., and ¾-in. thicknesses. Mahogany comes in a range of shades and grain patterns and takes stains well if you choose to alter the natural color. It's easy to work using both hand tools and machines and is especially good for carving and turning. This versatility makes Honduras mahogany one of my favorite woods. The fact that the Greene brothers used it extensively in their interior projects made it an obvious choice for this mantel. You may come across other species of mahogany. African mahogany is a suitable substitute for Honduras mahogany. Lauan and meranti are lower grades of mahogany; though less expensive, they don't have the visual character or working properties of Honduras mahogany.

ROUTING THE "CLOUD LIFT" MOTIF

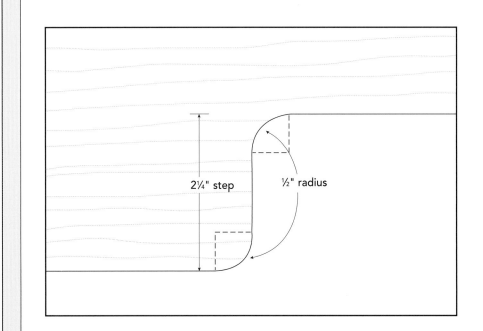

2¼" step   ½" radius

1. Trace the outline onto the workpiece from a template.
2. Cut the workpiece to within ⅛ in. of the outline on the bandsaw.
3. Tack or screw the template to the back of the workpiece and clamp everything securely to the workbench, making sure the edge of the workpiece clears the bench.
4. Using a pattern-routing bit, carefully set the rotating bit against the template and shape the parts to conform to the template.
5. Round over the corners of the parts using a ½-in.-radius bit on the shelf and brackets and a ⅜-in.-radius bit on the H-frame parts.

The "cloud lift" profile on the rail is roughsawn to within ⅛ in. of the outline, then trimmed to exact shape with a pattern-routing bit.

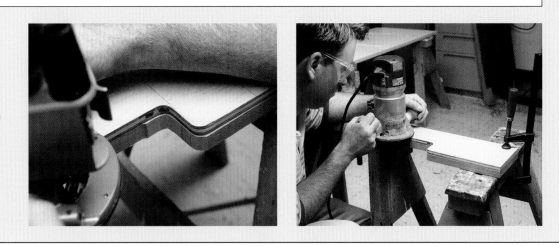

3. Shape the top corners of the stiles and the bottom edge of the rail to form the cloud lift motif details (see the drawing above).

4. Rout a ⅜-in.-diameter quarter-round profile along the outer edge of the stiles, but leave the inside edge crisp. Lightly round over the bottom edge of the rail by hand with sandpaper.

5. Lay out and cut biscuit joints to join the rail to the stiles, and dry-assemble the frame.

BUILDING THE OVERMANTEL

The overmantel is constructed as a self-contained frame-and-panel unit that is slipped between the H-frame stiles and held in place with a ¼-in.-thick spline. It consists of a pair of "cloud lift" rails set between 6-in.-wide stiles, supporting a ¼-in. mahogany plywood panel.

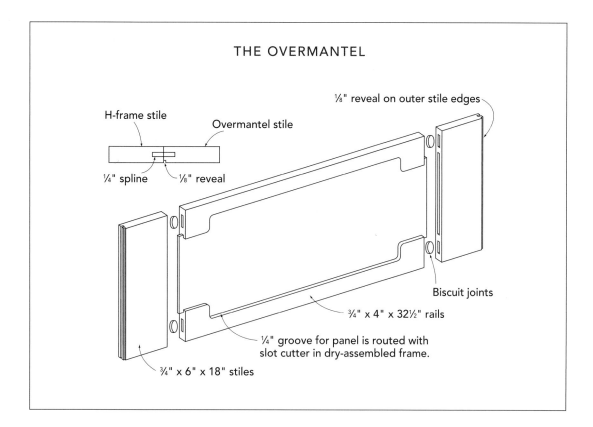

THE OVERMANTEL

H-frame stile

Overmantel stile

¼" spline ⅛" reveal

⅛" reveal on outer stile edges

Biscuit joints

¾" x 4" x 32½" rails

¼" groove for panel is routed with slot cutter in dry-assembled frame.

¾" x 6" x 18" stiles

Making the frame

1. Cut the overmantel rail and stile stock to width and length.

2. Shape the parts as shown in the drawing on the facing page.

3. Cut biscuit joints to join the frame parts.

4. Rout a groove along the inside of the frame parts using a ¼-in. slotting bit on the router. I did this with the frame dry-assembled, but you can groove the parts separately if you like— just be sure to stop the groove before running into the biscuit slots in the stiles.

Fitting the panel

1. With the frame dry-assembled, trace the inside of the frame onto a scrap piece of ply-wood to use as a template.

2. Cut out the template panel, and clean up the corners with a rasp and some sandpaper. The template should just fit inside the assembled frame, not inside the grooves.

3. Lay the template on the ¼-in. mahogany plywood panel. With the legs of a compass set ⅜ in. apart, trace around the template to project the actual panel dimensions and shape.

Make a template that fits inside the overmantel frame, then extend the shape by ⅜ in. using a scribing compass. The larger profile traced on the mahogany panel will fit into the overmantel frame groove.

This ⅜-in. extension fits into the groove routed in the frame parts.

4. Cut out the overmantel panel with a saber saw or on the bandsaw, and check the fit within the assembled frame.

Completing the overmantel

1. Before gluing up the overmantel, rout or saw a ¼-in.-wide groove along the outer edge of the overmantel stiles. This can be done with the same slotting bit used for the groove in the frame parts. Cut the same groove along the inside edge of each H-frame stile, from the top down to the intersection with the rail.

2. Cut a reveal in the face of the overmantel stiles. This small space breaks up the large expanse of wood and visually separates the overmantel from the outer stiles.

3. Sand the parts of the overmantel, then the frame around the panel.

MAKING THE STEPPED BRACKETS

To get a clean stepped shape on the brackets, and tight-fitting dadoes for the shelf, be sure your bracket stock is milled good and square to start.

This detail shows the thick mantel shelf supported by a notch in the bracket.

Plowing out the dadoes

1. Lay out the dado on the inside edge of each bracket.

2. Plow out the dado on the table saw. A dado blade will do the job more quickly, but making repeated cuts with a saw blade will get the job done as well.

3. Clean up the bottom of the dadoes with a wide chisel or a rabbet plane.

Shaping the bracket steps

1. Drill out the inside corners of the steps with a ⁷⁄₁₆-in. drill bit. It's a lot easier to create this corner with a drill than with any saw. A drill press is best, but if you use a hand drill, lay out the holes on both sides of the bracket and drill halfway from each face.

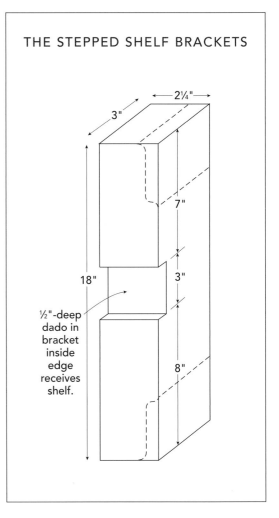

THE STEPPED SHELF BRACKETS

2¼"
3"
7"
3"
18"
8"
½"-deep dado in bracket inside edge receives shelf.

Cut the dado in the stepped bracket on the table saw by making repeated cuts between scribed lines indicating the thickness of the mantel shelf. The waste can be removed with a wide, sharp chisel or a shoulder plane.

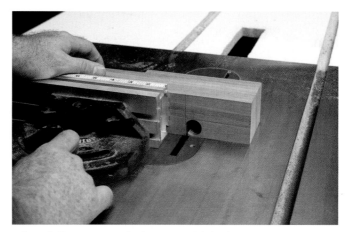

After drilling a 7/16-in. hole to establish the round corner, cut the short shoulder of the stepped bracket on the table saw. Don't cut too deep or you'll damage the rounded inside corner.

Cut the long shoulder on the bandsaw, using a rip fence as a guide. Clean up the cuts with a block plane, rasps and files, and sandpaper.

This detail shows the stepped bracket with square rosewood pegs.

2. Cut the short shoulders on the table saw. Be sure the blade is not so high that it cuts into the rounded corner.

3. Cut the long shoulder on the bandsaw.

4. Round over all the sharp edges and corners with a 1/4-in.-radius quarter-round bit, then sand the brackets to 220 grit.

MAKING THE SHELF

The mantel shelf is a single solid slab of 3-in.-thick mahogany that resembles the huge timbers often used in Greene and Greene inte-riors. I achieved a characteristic softness by rounding the edges and shaping the ends of the massive shelf.

1. Plane the shelf to thickness so it fits snugly in the bracket dadoes.

2. Rip the slab to 6 in. wide and crosscut it to length.

3. Shape the step on each end of the shelf using the same sequence used to shape the brackets: Drill the starter hole, cut the short shoulder on the table saw, and cut the long shoulder on the bandsaw.

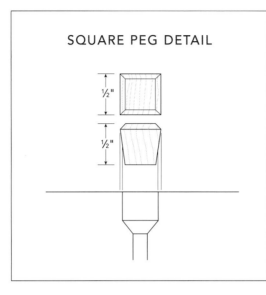

SQUARE PEG DETAIL

½"

½"

4. Cut the ½-in. by 2¼-in. notch on each end of the shelf (see the drawing on p. 119). This notch allows the shelf to interlock with the bracket better, and also has the rounded edge of the shelf ending neatly against the flat face of the bracket.

5. Saw a ½-in.-deep by ¾-in.-wide groove centered in the back edge of the shelf.

6. Round all the edges with a ¼-in.-radius router bit, and sand the shelf to 220 grit.

FINAL DETAILS AND ASSEMBLY

Cutting the square pegs

The contrasting square pegs are another well-known Greene and Greene design feature. This distinctive touch punctuates much of their work and gives it an exotic look. They are also functional—hiding the screws that hold the mantel together and against the wall. The pegs are tapered to be force-fit into round holes and left proud of the surrounding surface.

1. Predrill ⁷⁄₁₆-in. holes in the brackets and at the top and bottom of the H-frame stiles as indicated in the drawing on p. 118. (The exact location is not crucial.) Also drill a couple holes in a scrap piece of mahogany to test-fit the pegs.

2. Plane some rosewood stock to ½ in. square.

3. Crosscut the stock into convenient lengths—about 6 in. long is comfortable. Then chamfer around the four corners of each end. I used a shop-made jig and a block plane.

4. Taper each peg along its length. I made a simple bandsaw jig for this. Cut the pegs from the longer stock on the bandsaw. You should end up with a square tapered peg with a chamfered top that can be coaxed into a ⁷⁄₁₆-in.-diameter hole. As the peg is driven into the wood, the sides of the round hole yield to accept the increasing girth of the rosewood peg. When the base of the chamfer is flush with the work surface, the peg is home.

The ends of the ½-in.-square peg material are chamfered using a block plane and this holding jig.

This jig guides the peg stock on the bandsaw to shave a slight taper along the length of the peg.

The peg should be neatly chamfered and taper evenly before being cut free.

A tapered peg is set into a countersunk hole to cover a #8 by 1½-in. wood screw.

The peg should be driven into the hole until the base of the chamfer is flush with the wood surface.

Making the crown

For this simple crown, I used some 5/4 mahogany stock planed to 1 in. thick and ripped to 1¾ in. wide. When installed on the mantel, it provides a neat finishing touch with about ¾-in. overhang. The crown is prepared to resemble the breadboard end of a tabletop. The center section presents long grain to the viewer. The two 3-in. end pieces have their end grain facing out, and the splines joining everything together are made of rosewood.

1. Plane the parts to 1 in. thick.
2. Rip the long center strip to 1¾ in. Rip a 3-in.-wide strip for the two end pieces.

The mantel crown is really more of a simple but elegant cap. The "breadboard ends" and an exposed rosewood spline form a typical Greene and Greene detail.

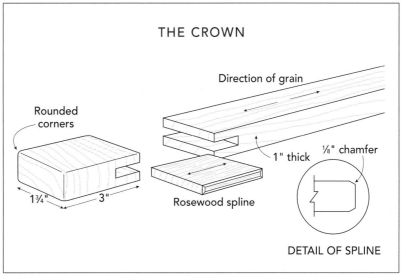

THE CROWN

Rounded corners

Direction of grain

⅛" chamfer

1" thick

1¾"   3"

Rosewood spline

DETAIL OF SPLINE

The H-frame is assembled with a single clamp. Check that the assembly is square, and attach a temporary brace to the back if you need to move the assembly.

Slide the completed overmantel assembly into position on the H-frame.

Position the assembled H-frame over the shelf cleat, then screw it to the fireplace wall.

3. Crosscut the center strip to length, but leave the two short end pieces uncut for now.
4. Rout a groove into the ends of the center strip, and into one edge of the wider strip; then crosscut the two short end pieces from the wider strip.
5. Plane the rosewood spline material so it fits the grooves snugly, then cut it to size. Using a block plane, chamfer all four corners along one long-grain side of each spline (see the drawing on p. 125).
6. Assemble and glue the ends onto the center section so the chamfer on the splines protrudes evenly.

Assembly and installation

1. Glue the main H-frame together.
2. Glue the splines into the grooves in the upper stiles, then slide the overmantel onto the splines. Note that there should be a ¾-in. gap between the overmantel and the H-frame rail for the wall-mounting cleat.
3. Screw a mounting cleat to the wall to receive the mantel. The cleat is located so it falls in the gap between the overmantel and the H-frame rail.
4. Position the assembled mantel on the wall cleat. Drive screws through the predrilled holes at the top and bottom of the stiles. Additional screws can be used if needed where they'll be hidden by the shelf brackets.
5. Screw the shelf brackets to the stiles. (Note: It may be easier to attach the brackets to the H-frame while it's lying flat, but I wanted to

determine the exact height of the shelf after the rest of the mantel was in place on the wall.)

6. Screw the ½-in. by ¾-in. shelf support strip to the H-frame rail.

7. Glue the shelf into the bracket dadoes. Add a few finish nails along the back edge, angled into the support strip, to lock the shelf in place.

8. Drive in the rosewood pegs.

9. Position the crown and secure it with a couple finish-head screws or sixpenny nails through predrilled holes.

Finishing the mantel

Greene and Greene woodwork isn't slick or glossy but instead exhibits a soft and natural glow. To imitate this attractive finish, I applied three coats of water-based polyurethane onto the natural mahogany. Each coat was brushed on, allowed to dry, then rubbed out with #000 steel wool. The final step was to polish the mantel with a clear furniture paste wax. This is a simple finish, but it achieves the appropriate low-key surface that complements the design perfectly.

An optional step is adding a glaze coat. I applied a light glaze made from a mixture of raw-umber Japan pigment and a standard glazing product.

Apply glue to the groove in the back of the shelf, then slip the shelf into the bracket dadoes and over the wall cleat. A couple of finish nails angled down through the back of the shelf and into the cleat will lock the shelf firmly while the glue sets.

Drive the square pegs into the countersunk holes, keeping the sides of the pegs square to the workpiece.

This detail of the overmantel shows the "cloud lift" pattern.

SKETCHBOOK: A SURVEY OF GREENE AND GREENE DESIGN ELEMENTS

Throughout their work, the Greene brothers employed a number of distinctive decorative devices that stemmed from their fascination with Japan. The Greenes' use of large beams, exposed joinery, protruding pegs, wrought-iron strapping, and braces were a fusion of Japanese design elements and the southern California landscape, climate, and lifestyle. This synthesis produced architecture and interiors of unique and timeless beauty.

These sketches are my "doodles," a way to familiarize myself with these unique details and to consider where and how they might be employed on my future mantel designs and in other woodworking adventures.

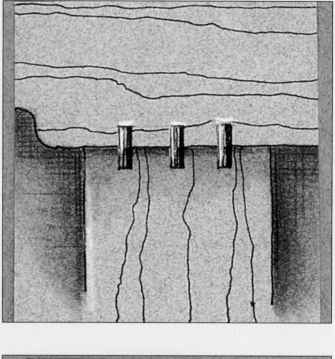

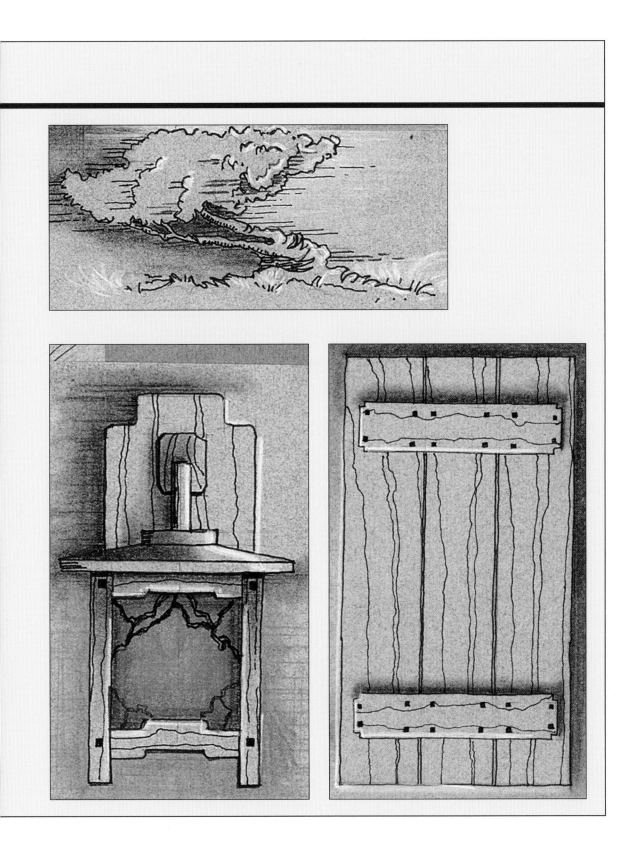

Art Deco Mantel

By the 1920s, the family was no longer tethered to the domestic fireplace. Now there was central heating, electric light, hot water on tap, the modern kitchen. The pace of life had accelerated, and people no longer planned their activities by the light of day or the heat of the hearth. The traditional fireplace became a sentimental fixture, a quaint reminder of simpler times. If there had to be a fireplace to augment the central heating source, it should be discreet, tucked behind a sculptural façade or surrounded by an architectural frame that diminished the attention previously conferred upon the traditional hearth.

Architecturally, the design of Art Deco buildings suggested movement, speed, and progress. This mantel, based on design features from that era, might have been found in a metropolitan high-rise apartment. It neatly frames the hearth, but in a playful, almost zany, manner. The all-over zigzag pattern is sculptural and exuberant, marking a clear break with traditional mantel designs. Although simple, the design isn't static or monotonous, but instead generates visual energy.

The mantel face is made up entirely of triangular sections (I call them "pyramid moldings"), laid horizontally and wrapping neatly around the corners. Each section is a simple L-shaped plywood assembly formed with a plain miter joint. Resting on top of unadorned plinths, each section is nailed to a plywood foundation, culminating in a square-edged mantel shelf.

Art Deco Mantel

CELEBRATING A ZEST FOR CHANGE AND NOVELTY that characterized the Art Deco period, this mantel is stripped of any traditional details. Instead, a simple V-shaped molding wraps the entire form like a crazy quilt. The resulting texture is inviting exactly because it's so regimented, but overall the mantel makes a bold architectural statement nonetheless.

A PLAN VIEW

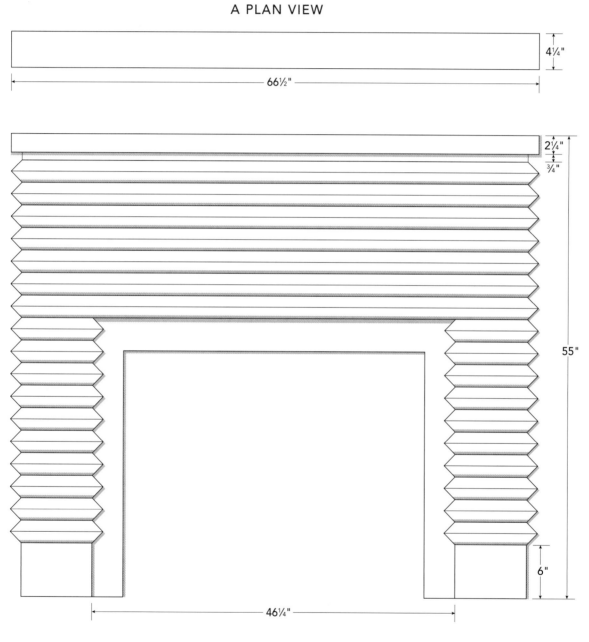

4¼"

66½"

2¼"

¾"

55"

6"

46¼"

FRONT VIEW

BUILDING THE MANTEL STEP-BY-STEP

This mantel goes together in typical sequence—construct a foundation; assemble the face pieces, the plinth blocks, and mantel shelf; then attach the elements to the foundation. The mantel can be installed in one finished piece.

CONSTRUCTING THE FOUNDATION

The pyramid moldings that make up the finished mantel are nailed to a simple plywood foundation. The foundation is 2½ in. deep, providing a modest amount of projection from the wall.

Building the foundation

1. Cut the foundation parts to size. I used ¾-in.-thick shop-grade birch plywood.
2. Cut biscuit slots to join the lintel to the pilasters, then glue up the face of the foundation. Make sure the assembly is square and flat.

The foundation is made of plywood—narrow ribs faced with plywood and screwed together. Lines mark the location of the pyramid moldings that will cover the foundation completely.

3. Lay the foundation face onto the plywood ribs and screw the parts together using countersunk #8 drywall screws. All the plywood and screws will be covered over with the pyramid molding.

CHOOSING MATERIALS

For this mantel I chose mahogany. Its deep, rich color makes the finished mantel suitable for the modern home, apartment, or city loft.

You could use solid mahogany stock to make the pyramid moldings on this mantel, but I used ¾-in.-thick plywood instead. With the benefits of uniform grain patterns and structural stability, plywood is ideal for this project. You can get the whole project from a single 4 by 8 sheet, with some to spare.

If I'd built this mantel from solid wood, the design would have required lots more work,

with questionable results. For instance, removing large amounts of wood from a squared piece of mahogany might cause the piece to warp or twist. Then aligning one piece to the next would be difficult at best.

Another problem using solid wood would be the natural disparity in color and grain between the layers of molding. The key to the success of this mantel design is a certain uniformity of color between the layers.

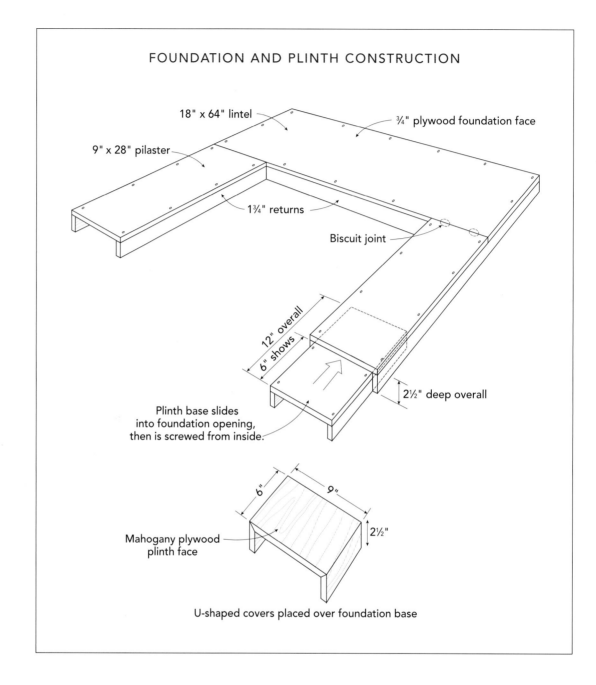

FOUNDATION AND PLINTH CONSTRUCTION

18" x 64" lintel

¾" plywood foundation face

9" x 28" pilaster

1¾" returns

Biscuit joint

12" overall

6" shows

2½" deep overall

Plinth base slides into foundation opening, then is screwed from inside.

6"

9"

2½"

Mahogany plywood plinth face

U-shaped covers placed over foundation base

Making the plinths

The plinths have two parts, the base and the face. The base assembly slides into the foundation. The face slides over the base.

1. First make the U-shaped plinth bases that fit within the foundation (see the drawing above), using ¾-in. plywood. Slip them into the foundation and screw them from behind so that 6 in. of the base is exposed.

2. Make a similar assembly for the plinth faces, but use mahogany plywood, and glue and miter the corners.

3. Depending on your site conditions, you can leave the plinth faces a little long in case they need to be scribed to the floor at installation. Also, don't attach the face plinths until you've applied the face molding to the foundation; the bottom piece may hang over slightly, in which case you can cut the face plinth to fit.

MAKING THE PYRAMID MOLDINGS

The pyramid moldings are simple L-shaped plywood assemblies that wrap around the foundation completely. Assemble them in long lengths and then cut them to fit.

Mitering the plywood

When mitering plywood, I do the job on the table saw in a way that cuts through the full thickness of the plywood backing material but maintains the full thickness of the face veneer. The method involves no special machinery, bits, or cutters and is completely performed on the table saw. (See the drawing on p. 69.)

1. Rip all the mahogany plywood strips to 2¼ in. wide. This is the finished width, and the face doesn't get recut when you're cutting the miters.

2. Attach an auxiliary fence onto the regular table saw fence. Mark the exact thickness of the ¾-in. mahogany plywood material on the fence.

3. Set the table saw blade at 45 degrees, make a test cut to check the angle, then lower the blade to about ¾ in. high.

4. Adjust the fence so the blade will cut into the plywood fence just below the pencil line indicating the exact thickness of the mahogany plywood, then rip both edges of the plywood strips.

5. In the event that a section of a strip lifts off the table, just run it over the angled blade

A strip of the ¾-in. mahogany plywood is laid flat on the saw table, and its exact thickness is marked onto the auxiliary fence.

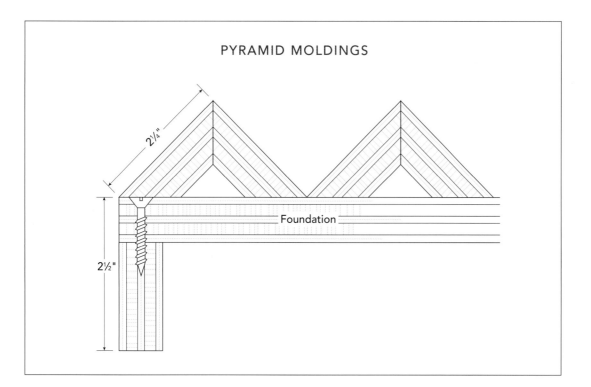

PYRAMID MOLDINGS

2¼"

2½"

Foundation

After setting the blade angle to 45 degrees (and making a test cut), the blade is raised to cut into the auxiliary fence, just below the thickness line.

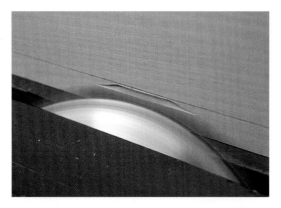

This detail shows the pyramid molding resting on the plinth base.

The edge of a mahogany plywood strip is cut at 45 degrees without cutting through the face veneer.

Position two of the mitered strips in the simple assembly jig and nail them together.

again. There is no need to adjust or reposition the blade or fence. The only material that will be removed is the excess—and the face veneer will remain intact.

Assembling the pyramid moldings

1. Sand the face of the mitered stock. Since it's a lot easier to sand the material when it is flat, carefully go over it with 120-grit sandpaper, then 180-grit, before assembly. I constructed a jig to support the sander and prevent "sand-throughs" at the delicate edges.

2. Make an assembly jig, also shown in the drawing on the facing page.

3. Apply glue to the miters and load the pairs into the assembly jig.

4. Nail the assembled pieces together with a pin nailer, spacing the nails 6 in. apart. Here you must work quickly because the glue will slightly swell the surface of the miter, making a tight joint difficult to obtain if you dawdle.

5. Wipe down each assembled pyramid molding strip with a warm wet rag to remove any glue. Before the pyramid molding strips dry, check to see whether any portion of the joint is open. If the lighter-colored core of the ply-

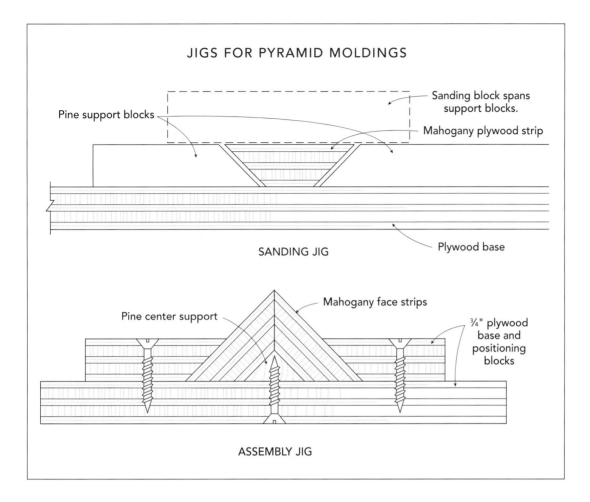

JIGS FOR PYRAMID MOLDINGS

Sanding block spans support blocks.

Pine support blocks

Mahogany plywood strip

SANDING JIG

Plywood base

Pine center support

Mahogany face strips

¾" plywood base and positioning blocks

ASSEMBLY JIG

wood is visible, take a screwdriver or round burnisher and gently rub it across the area to close any small gap.

6. When the assembled triangles are dry, sand them with 220-grit sandpaper.

ASSEMBLING THE MANTEL

Cutting the corner miters

Since the pyramid molding wraps around the foundation and returns to the wall, it must be mitered at each end. I made a simple L-shaped table saw fence to support the miter cut, as well as a bandsaw jig to guide the crosscutting of the short mitered returns.

1. Miter-cut one end of all the front pyramid pieces. Use an L-shaped jig to support the workpiece.

Cut the end miters on the table saw with the aid of an L-shaped fence.

The bandsaw offers a safe way to cut the short miter returns. This jig ensures that each piece is cut square and to the same length.

The compound miter cut should be clean and gap-free.

The completed strips are laid onto the foundation to establish their exact location and precise spacing.

2. Position the pieces on the foundation with the back of the already-cut miter aligned with the edge of the foundation, and mark the second miter cut on the back of the molding stock.

3. Set a stop on the saw fence to regulate the length of each piece. Cut all the long pieces for the top section of the mantel, then move the stop and cut all the short pieces for the lower sections.

4. For the short returns, miter both ends of an 8-in.-long piece on the table saw. Then use a holding jig to make the square cut on the bandsaw. Don't try to cut the short returns on the table saw—it's unsafe!

5. Check that the joints are tight and square when two end mitered pieces are held together. Careful test cuts and the use of shop-made jigs made these crucial cuts clean and resulted in gap-free miter joints.

Attaching the pyramid molding

1. Draw lines on the foundation and lay out the assembled moldings to check the overall fit. You need to end up with a full molding at the top and bottom.

2. Start from the bottom and attach one layer at a time. Nail the pyramid moldings on with a brad nailer.

3. Attach all the face pieces first, then go back and nail on all the returns.

4. After each layer is attached to the foundation, clean up any excess glue. When the assembly of the mantel is complete, carefully sand the miters with 220-grit sandpaper.

Making the mantel shelf

The mantel shelf is made of three layers of ¾-in. veneered plywood, with a fourth layer recessed to form a reveal between the shelf and the face moldings. For an Art Deco touch, I veneered the shelf's front and side edges with the grain running vertically. (Alternatively, you could miter pieces of plywood with the grain running vertically to form the edge.) The shelf is 2¼ in. thick and is flush with the tips of the pyramid molding to provide some visual balance.

Glue and nail the molding strips to the mantel foundation.

1. Cut the shelf parts slightly oversize. The top is mahogany, but the other two pieces can be any ¾-in. material.

2. Nail and glue the three shelf pieces together. Note that the middle piece is narrower to create a gap at the back for a mounting strip.

3. Recut the assembled shelf to finished size.

4. Veneer the front and side edges with vertical-grain mahogany veneer.

5. Make the reveal strip, which sits flush with the foundation assembly, and veneer its front and ends with mahogany.

FINISHING THE MANTEL

I stained the entire mantel in a light, warm walnut stain, applied with a brush and rag, and allowed it to dry overnight. The next day, I applied a thinned-out coat of blond shellac to seal the stain. Then I restained the mantel to achieve a uniform color. To suggest some subtle aging, I applied a heavier coat at the base of the mantel and on the returns.

Adding highlights

To give some interest to the mahogany and create highlights, I applied a golden yellow wash, composed of Japan color suspended in

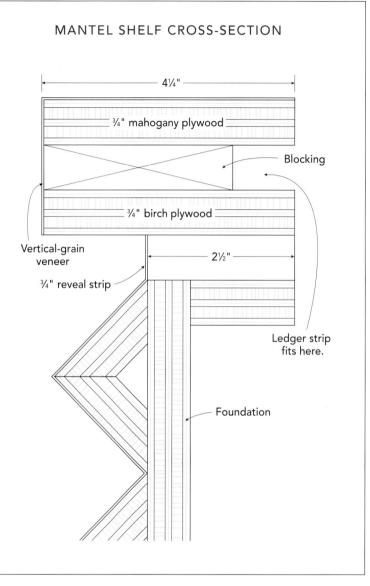

MANTEL SHELF CROSS-SECTION

4¼"

¾" mahogany plywood

Blocking

¾" birch plywood

Vertical-grain veneer

¾" reveal strip

2½"

Ledger strip fits here.

Foundation

Tip: Be sure to lay out the molding on the foundation before attaching it, to ensure that you don't end up with a partial molding at top or bottom. For a slight discrepancy, you can rip the top or bottom run of moldings, or adjust the size of the face plinth. For a larger discrepancy, you may have to cut the legs of the foundation, which is easily done with a jigsaw.

ALTERNATIVE FACE-MOLDING PROFILES

The Art Deco period was swept up in exotic themes executed with modern production methods and materials. Many design motifs were composed of angular geometric patterns. And these motifs were often arranged in repeating patterns or bands. Even flowers and other natural forms were depicted in shallow relief or simple rectilinear outlines. And furniture of the period was designed with great bulky curves, devoid of any fussy ornament, that suggested bold sweeping change, luxury, and progress. When veneer was used on furniture and architectural interiors, it was arranged in vivid patterns, or stained in unusual colors, then accented with ivory, onyx, or brass. The designers of the period rejected Romanticism, with its asymmetrical free-flowing arrangements taken from nature, and embraced a clean, modern aesthetic.

In addition to the angular pyramid molding used in this mantel, I've illustrated two other suitable patterns for the face of this mantel. Either one could be used successfully on the same mantel design, but each would produce a distinctly different result.

The square-edged treatment would create a stronger light-and-shadow effect and could be

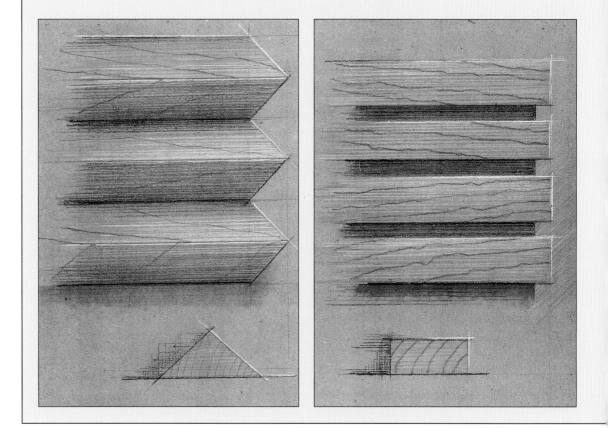

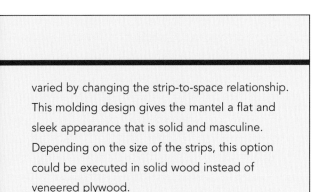

varied by changing the strip-to-space relationship. This molding design gives the mantel a flat and sleek appearance that is solid and masculine. Depending on the size of the strips, this option could be executed in solid wood instead of veneered plywood.

The other is a half-round profile which produces a subtle shadow, and is probably more commonly associated with the Art Deco period. In this design, like the pyramid molding, the half-round molding strips are butted together without any space, and the result is a softer, less jarring look.

turpentine and applied with a rag. This treatment deposits a thin film over the mahogany, giving it a golden tone. When the golden wash was dry, I applied two more coats of shellac.

Later, to accentuate the heavy zigzag design, I also applied an antique glaze. This treatment comes up frequently in this book because it most effectively reproduces the appearance of an antique mantel. I don't intend to fool anyone, but a mantel with an old appearance will fit into a historic room a little better—its lack of age won't be quite so noticeable.

INSTALLING THE MANTEL

On a mantel like this, hiding screw holes can be difficult. You wouldn't want anything to detract from its clean appearance. And I wanted to keep everyone guessing as to how it was attached to the wall.

Behind the hollow foundation, I placed locator-support blocks. I used three blocks across the top and one at the bottom of each pilaster. These can be secured to the wall with masonry screws, hollow-wall fasteners, or wood screws driven into the wall studs.

Once the mounting blocks are attached to the wall, position the completed mantel, minus the mantel shelf, over the blocks.

1. Check the mantel for level and plumb, and then screw through the top of the plywood foundation and through the plinth bases into the blocks.

2. Fit the plinth bases, scribe if necessary, nail them onto the plinth bases, and fill the nail holes with a wax filler.

3. Nail down the reveal strip on top of the foundation.

4. To attach the mantel shelf, screw a mounting block to the wall so it will align with the groove in the back of the shelf. Position the shelf over the mounting block, drive a couple finish-head screws down through the back edge of the shelf into the mounting block, and again fill these holes with wax.

Contemporary Architectural Mantel

For this modern interior, I knew that a fireplace mantel with traditional features just wouldn't do. Instead I imagined something large and sculptural. Something that was part of the architecture, yet springing from it, almost attempting to break free from the wall. Abandoning the typical mantel shelf, I designed this monolithic structure that juts cockily from the wall, suggesting (to me, anyway) a puffed-up turkey breast. Triangular in cross section, it looks a bit like a range hood over the fireplace. Down lights in the underside of the "box" cast an artful glow where a mantel shelf would typically be found. You might call it the anti-mantel, up to that point.

Perhaps a bit too austere, though. So with a wry nod to tradition, I added the red shelf. The shelf straddles the apex of the structure, piercing it like a thick arrow, and provides two bright surfaces for displaying small pottery or other items.

Though large overall, this is an uncomplicated design. Triangular horizontal struts define the shape of the structure, which is sheathed on the two vertical faces with ¾-in. plywood. The back is open, providing area for simple French cleats to hang the whole thing on the wall. The front corner is 90 degrees, which makes that end of things easy. The shelf parts are cleated to the face after assembly.

Contemporary Architectural Mantel

THIS MANTEL IS DESIGNED TO BE PART OF THE ROOM unlike any other mantel in the book. The adjacent walls are paneled with simple oak plywood, and the large mantel box is built of the same material so it appears to be part of the wall. The angular design of the mantel makes for a few challenges, but the construction is surprisingly basic.

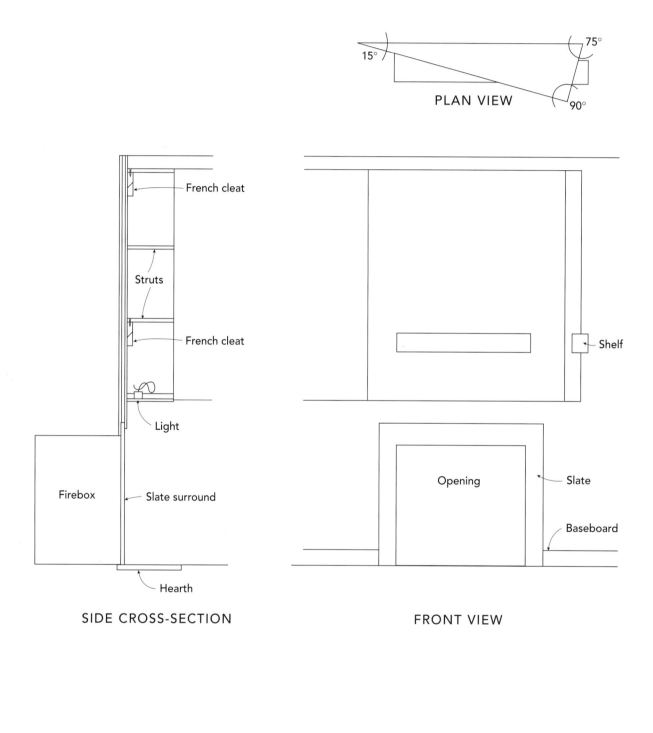

PLAN VIEW

15°   75°   90°

SIDE CROSS-SECTION

French cleat

Struts

French cleat

Light

Firebox

Slate surround

Hearth

FRONT VIEW

Shelf

Opening

Slate

Baseboard

BUILDING THE MANTEL STEP-BY-STEP

There are only three elements to this mantel, so the sequence of construction is simple. First build the big overmantel box, then make the two shelf pieces. Prefinish the parts separately, then install them. To ensure that everything would come together properly, I worked from a full-sized master drawing. Every piece, as it came off the saw, was measured and checked against the drawing.

MAKING THE OVERMANTEL

The overmantel is composed of four horizontal struts and the two vertical face pieces that form the exterior of the structure. The left edge of the main face piece is cut at a very acute angle; I cut the angle on a smaller, more manageable piece first, then joined it to the larger piece with supporting battens in the rear.

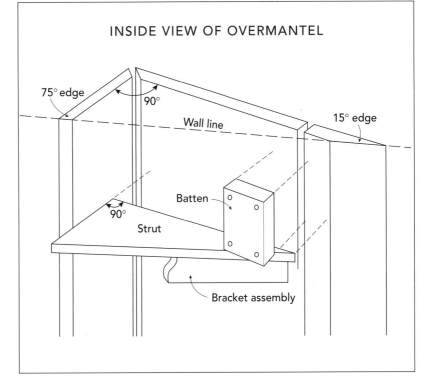

INSIDE VIEW OF OVERMANTEL

75° edge

90°

Wall line

15° edge

Batten

90°

Strut

Bracket assembly

CHOOSING MATERIALS

Because of the size of the overmantel structure, I was concerned about its weight, stability, and construction. Veneer-core plywood was a logical choice to address those concerns. Another consideration was the appearance of the piece. It was going to cover a large expanse of wall and project into the room, drawing lots of attention. It had better look good.

I wanted a handsome straight-grained wood without a pronounced pattern. Riftsawn red oak fit the bill perfectly. The wood has texture but little pattern, for a subdued wallpaper-like look. I wanted to use the same material on the surrounding wall, and with riftsawn oak I could be sure that the surfaces on both the overmantel and the paneled wall would match perfectly. It also takes stain very evenly if you want to go for a darker shade.

Riftsawn oak is readily available from hardwood plywood suppliers, though you may have to special-order it from a general-purpose lumberyard or home center.

Cutting the face panels

One edge of each face panel is cut at a 45-degree miter to form the outside corner. The other edge of each is angled to meet the wall. The narrow piece gets a 75-degree cut, which is a routine cut, but the wide piece gets a knife-edge rip of only 15 degrees. That takes a little more ingenuity.

1. Rip the narrow side panel to size. First rip the 75-degree back edge. Then set the blade to 45 degrees and rip the front edge. I use a simple system for ripping 45-degree miters in plywood that protects the fragile edge and prevents making a miscut. (See p. 69 for a detailed description of this technique.)

2. To cut the knife edge on the wide face panel, the panel needs to be passed vertically across the blade set at a 15-degree angle, and with the fence to the left side of the blade. This is impossible with a wide panel, so I first ripped an 8-in. section from the panel and cut the knife edge on it. Then I reglued this section back onto the wide panel, using biscuit joints for alignment and battens across the back for strength. Finally, I ripped the 45-degree edge of the front panel.

3. Crosscut the two face panels to length.

Cutting the struts

The top and bottom of the overmantel are identical to the intermediate struts, so you need a total of four struts.

1. Make a template for the triangular struts from ¼-in. plywood. Start with a piece of square material, and use the 90-degree corner as the front corner of the strut. That way you're making only one cut to form the back edge of the strut.

2. Cut the strut template on the bandsaw or jigsaw, then straighten the back edge cut on the jointer or by planing carefully to the layout line.

3. Cut out the two struts (and the top and bottom, which are identical) from ¾-in.-thick plywood, and trim the back edge of each to the template using a pattern-routing or flush-trimming router bit.

Assembling the overmantel box

1. Lay the face panel upside down on sawhorses and draw layout lines for the two intermediate struts.

2. Make the bracket assemblies shown in the drawing on p. 145 and the photo at left, and attach one to each of the four struts. Position the top, bottom, and two intermediate struts, and screw the brackets to the face panel with

---

CUTTING LARGE PANELS

Many table saws are limited in their cutting capacity when it comes to large panels, like the main face piece on this overmantel. Here's how I get straight square cuts when the panels are too big or just too unwieldy for my table saw. First, I take advantage of factory edges wherever possible. They're almost always good and straight, and if I can use two adjacent edges and the corner is square, that's even better. Check the edges with a straightedge and the corners with a framing square. Next, I lay out my cut lines with a 4-ft. drywall T-square. Using a saber saw, I'll cut about ⅟₁₆ in. outside the line. Finally, I clamp a straightedge on the cut line and use a router with a flush-trimming bit to trim right to the line.

First screw a bracket assembly to each strut, then screw the bracket to the face panel.

Join the mitered corner of the overmantel with narrow-crown staples judiciously placed along the corner from both sides.

To fill inevitable gaps, press glue into openings along the miter, and wipe away the excess with a damp cloth.

Then press a smooth wood block along the edge from both sides, gently closing any gaps before the glue sets.

1¼-in. drywall screws. Also, drive a screw through a predrilled hole about 2 in. from the pointed end of each strut.

3. Glue and nail square strips to the inside of the face panel along the inside edge of the miter. This will provide additional gluing and nailing surface for the main corner joint.

4. Apply glue to the corner miters, position the panels vertically, and nail the main corner together.

5. Clean up the mitered corner. You'll probably have some slight gaps along the miter, but they are easily filled with glue right after assembly. Rubbing a wood block along the corner will help "burnish" the gaps closed. When the glue has dried, sand the corner and break the sharp edge consistently along its length.

Adding lights

For a little extra drama, I installed small (20-watt) halogen lights to the underside of the overmantel. This would shed a soft curtain of light onto the slate hearth. The entire system was purchased as a kit and came complete

Low-watt halogen lights fit easily into drilled holes.

The clear directions provided with the light kit made installation quick and simple.

This detail shows applied oak paneling at the corner of the fireplace surround.

with three lights, wiring, on-off switch, and transformer. When the wiring was completed and the lights were installed, I located the transformer on the inside of the overmantel where it could easily be plugged into the wall receptacle, just before the overmantel was hung on the wall. When the installation of the overmantel was completed, the lights would be controlled from a switch on the wall.

1. Drill the mounting holes for the lights in the bottom of the overmantel. A hole saw works great here.

2. Attach the lights to the transformer and mount the transformer to the inside wall of the overmantel.

Finishing the overmantel

The straight grain and even surface of the red-oak plywood made this part of the job easy. To achieve a smooth satin finish that would survive mild abuse and clean up easily, I chose a water-based polyurethane. First, I applied a sealer coat with a foam-rubber roller, followed by two more light coats. I sanded between each coat with 150-grit sandpaper. The final coat was rubbed out with #000 steel wool, and waxed with an amber paste wax. Since I was covering the walls with the same red-oak panels, I finished them at the same time.

MAKING THE SHELF

The two-part mantel shelf is made from scraps of plywood glued and nailed together, then primed and coated with red spray paint.

Cutting and assembling the parts

The drawing on p. 150 shows the overall dimensions of the shelf boxes. The final dimensions, and how you configure the parts, are less important than the angles, which are complementary to the angles of the overmantel.

1. Determine the size of each part, as well as the angled cuts. Note that the back of each assembly is open to receive a mounting cleat.

2. Cut the parts to size on the table saw.

3. Assemble the two shelf boxes with glue and narrow-crown staples (or #6 finish nails).

Finishing the shelf

1. Fill the staple or nail holes with a latex filler, then sand the shelf boxes. Repeat this process, using finer sandpaper each time, until the surfaces are very smooth and have no noticeable flaws.

2. Spray or brush the shelf boxes with a primer coat of paint. After the paint dries, thoroughly scrape off any bumps with a single-edged razor blade, then fill in any remaining flaws.

3. Spray the boxes with enamel (I used Rust-Oleum®). Repeat the sand-and-patch sequence if there are any flaws after the first coat.

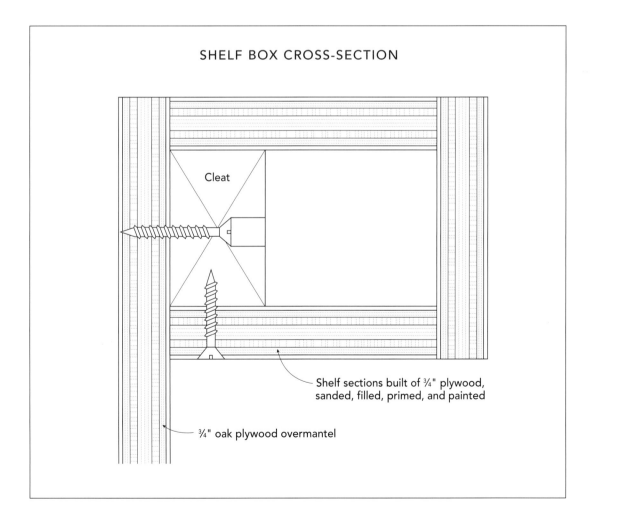

SHELF BOX CROSS-SECTION

Cleat

Shelf sections built of ¾" plywood, sanded, filled, primed, and painted

¾" oak plywood overmantel

THE SHELF BOXES

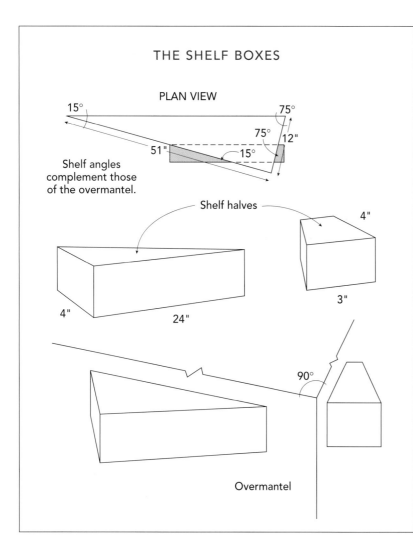

PLAN VIEW

15°   75°

75°

12"

51"   15°

Shelf angles complement those of the overmantel.

Shelf halves

4"

4"   24"   3"

90°

Overmantel

The initial coats of primer and paint are carefully sanded to remove any bumps and flaws.

4. Wet-sand the painted shelf boxes, starting with 320-grit and working up to 600-grit wet/dry paper for an ultra-smooth finish Finally, spray on two wet coats of satin-finish barn red enamel paint, and set the sections aside to dry thoroughly.

INSTALLING THE MANTEL

Paneling the fireplace wall

I checked the fireplace and adjoining walls for plumb and flatness and decided that the ¼-in. oak plywood panels could be attached to ¾-in. furring strips. However, on the fireplace wall, I had to double up the strips to clear the gray slate surrounding the firebox opening. I also had to cut an opening in the paneling for the switch that would control the lights. The base molding was painted to match the grey slate

The ¼-in. plywood is ripped into 24-in.-wide panels before being plumbed and stapled to furring strips.

surround, and set flush with the furring strips on the wall and the slate surround. The paneling, when attached to the furring strips, was placed to overlap the base molding and the slate surround. This gives the wall a clean, sleek and seamless look.

In order to make the wide sheets easier to handle, I ripped them in half, then attached them to the wall with a pneumatic staple gun. By holding the gun horizontally, the width of the small staple ran along the grain, making it easier to hide with wood filler.

Hanging the overmantel

1. Cut the strips of ¾-in. plywood for the French cleats (four 3-in.-wide pieces are needed).

2. Rip one edge on each piece at 45 degrees.

This panel was cut out for a wall outlet that provided electricity for the three small lights located in the bottom panel of the overmantel.

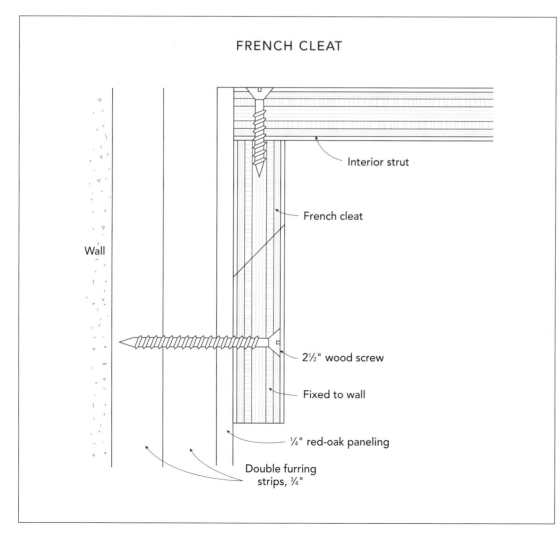

FRENCH CLEAT

Interior strut

French cleat

Wall

2½" wood screw

Fixed to wall

¼" red-oak paneling

Double furring strips, ¾"

Attach the French cleats to the wall with 2½-in.-long wood screws.

3. Screw two of the French cleat pieces to the overmantel, one under the top, and one under the lower intermediate strut (see the drawing on p. 144).

4. Once the paneling is installed, locate the two French cleats with level lines on the wall, then screw them to the wall. I used #8 by 2½-in. screws to make sure I hit the furring strips behind the paneling.

5. Position the overmantel just below its intended location on the wall. If you have lighting installed, this would be the time to thread the wiring. Then hoist the overmantel onto the wall-hung cleats. The whole process took me about 10 minutes. The best thing about using French cleats was that I could easily adjust the position and level of the overmantel by just relocating the wall cleats, then rehanging the overmantel.

Hanging the shelf sections

1. Mark out the location of the two shelf boxes on the wall.

2. Screw cleats to the wall to receive the shelf boxes. Use scraps of 2-by material or doubled up ¾-in. plywood. The cleats should be the same width as the opening in the back of the shelf boxes. Make sure the cleats are level.

3. Slip the shelf boxes over the cleats, then lock them in place them with 1¼-in. screws from underneath.

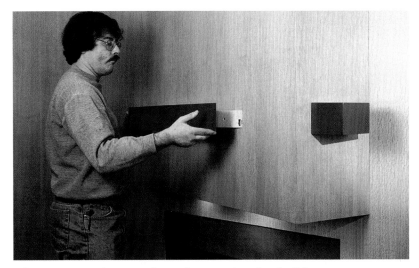

After screwing cleats to the wall to receive the shelf boxes, slip the boxes over the cleats.

Set the shelf sections in position on the cleats, and screw through the bottom into the cleat with 1¼-in. wood screws.

This detail of the smaller shelf box shows it hanging seamlessly from the overmantel wall.

ALTERNATIVE CONCEPTIONS FOR A CONTEMPORARY MANTEL

A firebox is a cavity cut into the wall, a negative space. Sometimes the space is filled with fire and light; other times it is a cold black hole. In either case the firebox tends to draw a viewer's eye away from the rest of the room and give nothing back. This mantel wrests the viewer's attention from the fireplace opening and focuses it on the wall above. There's a visual tug-of-war created by the positive and negative elements of the design.

These sketches explore this concept of positive and negative elements in a mantel. The designs are comprised of large architectural pieces that obviously abandon the symmetry of a traditional mantel and replace it with something sculptural and off-centered, something provocative.

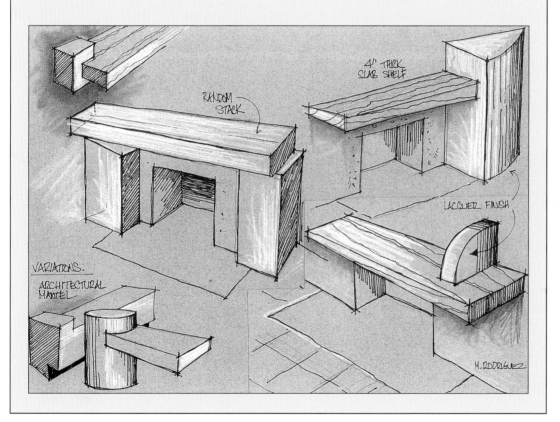